NAME:

_____

CLASS:

_____

Children often struggle with math. Of course, math is one of the most common and vast subject that children have difficulty with. However, it can be easier with the adequate effort, training and practice. For those 6th and 7th grade children who require extra practice to be good at math, **the Fast Math Success Workbook Grade 6-7** is the right decision. It provides variety of math topics aligned with common core standards with a lot of math problems.

This book provides sequential topics from foundations to harder topics because children can only understand advanced topics if they have completely grasped the basic concepts. For example, to solve an algebra or a fraction problem, children must know the basic addition, subtraction, multiplication, division and numeration.

Therefore, the best way to master math is to practice regularly and learn from the mistakes. This book provides practice worksheets on important math topics carefully selected for children to help advance in math. After completing this book, children will be able to recognize, understand, provide solution to a math problem. Moreover, this book will also build confidence for school tests and exams which are often unknown and create trouble for children.

We have provided solutions to all the practice problem given in the book; therefore, if you feel difficulty, just go to back of the book, locate the answer, and try to understand why this is the right answer and what you must do to come up with that solution. And once you understand the mistake, master the correct steps to solve the problem, and never repeat the mistake. **This is the secret way to Fast Math Success.**

This book can be used in classroom for math activity, extra practice, homework, homeschooling, and for teacher companion for non-prep guide.

Contents

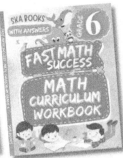

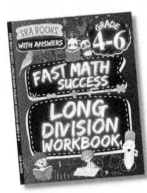

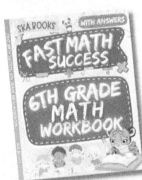

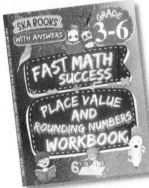

Name:
_____
Date:
__/__/____
Time Taken:
_____ Min

Multiplication with Whole Numbers

Multiply.

① $\frac{2}{4}$ of 2 = _____

② $5 \times \frac{7}{8}$ = _____

③ $\frac{2}{3}$ of 4 = _____

④ $6 \times \frac{3}{5}$ = _____

⑤ $\frac{1}{6}$ of 2 = _____

⑥ $\frac{2}{5}$ of 7 = _____

⑦ $\frac{3}{4}$ of 2 = _____

⑧ $\frac{3}{8}$ of 4 = _____

⑨ $2 \times \frac{2}{3}$ = _____

⑩ $5 \times \frac{1}{6}$ = _____

⑪ $5 \times \frac{2}{5}$ = _____

⑫ $\frac{2}{4}$ of 1 = _____

⑬ $4 \times \frac{1}{3}$ = _____

⑭ $7 \times \frac{4}{6}$ = _____

⑮ $\frac{6}{8}$ of 2 = _____

⑯ $\frac{5}{8}$ of 2 = _____

⑰ $9 \times \frac{1}{4}$ = _____

⑱ $\frac{3}{5}$ of 5 = _____

⑲ $\frac{1}{3}$ of 3 = _____

⑳ $1 \times \frac{1}{6}$ = _____

Fast Math Success Workbook Grade 6-7

Multiplication with Whole Numbers

Multiply.

(21) $\frac{3}{4}$ of 5 = _____

(22) $4 \times \frac{2}{3}$ = _____

(23) $\frac{1}{6}$ of 5 = _____

(24) $\frac{6}{8}$ of 2 = _____

(25) $\frac{3}{5}$ of 3 = _____

(26) $\frac{2}{3}$ of 5 = _____

(27) $\frac{1}{4}$ of 3 = _____

(28) $9 \times \frac{6}{8}$ = _____

(29) $6 \times \frac{3}{6}$ = _____

(30) $\frac{1}{5}$ of 4 = _____

(31) $2 \times \frac{2}{3}$ = _____

(32) $\frac{1}{4}$ of 7 = _____

(33) $\frac{3}{6}$ of 1 = _____

(34) $5 \times \frac{1}{5}$ = _____

(35) $3 \times \frac{5}{8}$ = _____

(36) $2 \times \frac{1}{3}$ = _____

(37) $\frac{2}{4}$ of 5 = _____

(38) $8 \times \frac{3}{5}$ = _____

(39) $2 \times \frac{2}{6}$ = _____

(40) $4 \times \frac{2}{8}$ = _____

Multiplication with Whole Numbers

Multiply.

(41) $1 \times \frac{1}{8} =$ _____

(42) $\frac{2}{4}$ of $1 =$ _____

(43) $4 \times \frac{2}{3} =$ _____

(44) $2 \times \frac{2}{5} =$ _____

(45) $\frac{3}{6}$ of $2 =$ _____

(46) $1 \times \frac{2}{5} =$ _____

(47) $4 \times \frac{3}{4} =$ _____

(48) $\frac{7}{8}$ of $5 =$ _____

(49) $2 \times \frac{2}{3} =$ _____

(50) $\frac{3}{6}$ of $8 =$ _____

(51) $3 \times \frac{5}{8} =$ _____

(52) $2 \times \frac{2}{3} =$ _____

(53) $\frac{2}{4}$ of $2 =$ _____

(54) $\frac{4}{5}$ of $2 =$ _____

(55) $2 \times \frac{4}{6} =$ _____

(56) $\frac{2}{5}$ of $2 =$ _____

(57) $1 \times \frac{1}{4} =$ _____

(58) $\frac{4}{6}$ of $5 =$ _____

(59) $1 \times \frac{1}{3} =$ _____

(60) $6 \times \frac{1}{8} =$ _____

Division with Whole Numbers

Divide.

(61) $\frac{2}{4} \div 8 =$ _____

(62) $\frac{2}{3} \div 4 =$ _____

(63) $\frac{3}{5} \div 9 =$ _____

(64) $\frac{4}{6} \div 9 =$ _____

(65) $\frac{1}{8} \div 5 =$ _____

(66) $\frac{4}{6} \div 4 =$ _____

(67) $\frac{3}{5} \div 8 =$ _____

(68) $\frac{2}{8} \div 9 =$ _____

(69) $\frac{1}{4} \div 7 =$ _____

(70) $\frac{2}{3} \div 3 =$ _____

(71) $\frac{4}{5} \div 4 =$ _____

(72) $\frac{3}{6} \div 6 =$ _____

(73) $\frac{2}{3} \div 8 =$ _____

(74) $\frac{3}{4} \div 9 =$ _____

(75) $\frac{4}{8} \div 7 =$ _____

(76) $\frac{2}{3} \div 1 =$ _____

(77) $\frac{2}{4} \div 4 =$ _____

(78) $\frac{1}{5} \div 4 =$ _____

(79) $\frac{2}{6} \div 8 =$ _____

(80) $\frac{5}{8} \div 7 =$ _____

Division with Whole Numbers

Divide.

(81) $\frac{4}{6} \div 3 =$ _____

(82) $\frac{5}{8} \div 7 =$ _____

(83) $\frac{1}{5} \div 6 =$ _____

(84) $\frac{1}{4} \div 2 =$ _____

(85) $\frac{2}{3} \div 5 =$ _____

(86) $\frac{4}{6} \div 7 =$ _____

(87) $\frac{1}{3} \div 3 =$ _____

(88) $\frac{4}{5} \div 3 =$ _____

(89) $\frac{3}{4} \div 9 =$ _____

(90) $\frac{3}{8} \div 6 =$ _____

(91) $\frac{1}{5} \div 7 =$ _____

(92) $\frac{2}{3} \div 1 =$ _____

(93) $\frac{3}{6} \div 9 =$ _____

(94) $\frac{3}{8} \div 4 =$ _____

(95) $\frac{3}{4} \div 2 =$ _____

(96) $\frac{2}{5} \div 4 =$ _____

(97) $\frac{6}{8} \div 6 =$ _____

(98) $\frac{4}{6} \div 4 =$ _____

(99) $\frac{3}{4} \div 3 =$ _____

(100) $\frac{2}{3} \div 9 =$ _____

Name:

_ _ _ _ _ _ _ _ _ _ _ _

Date:

_ _ / _ _ / _ _ _ _

Time Taken:

_ _ _ _ Min

Division with Whole Numbers

Divide.

(101) $\frac{2}{3} \div 4 =$ _____

(102) $\frac{3}{4} \div 3 =$ _____

(103) $\frac{2}{8} \div 8 =$ _____

(104) $\frac{1}{5} \div 5 =$ _____

(105) $\frac{3}{6} \div 3 =$ _____

(106) $\frac{1}{8} \div 3 =$ _____

(107) $\frac{2}{4} \div 6 =$ _____

(108) $\frac{1}{6} \div 9 =$ _____

(109) $\frac{2}{3} \div 6 =$ _____

(110) $\frac{1}{5} \div 7 =$ _____

(111) $\frac{5}{6} \div 8 =$ _____

(112) $\frac{6}{8} \div 2 =$ _____

(113) $\frac{1}{3} \div 4 =$ _____

(114) $\frac{2}{4} \div 2 =$ _____

(115) $\frac{3}{5} \div 1 =$ _____

(116) $\frac{1}{8} \div 6 =$ _____

(117) $\frac{1}{3} \div 8 =$ _____

(118) $\frac{1}{4} \div 8 =$ _____

(119) $\frac{1}{6} \div 9 =$ _____

(120) $\frac{3}{5} \div 1 =$ _____

Mixed Fractions - Multiplication

Calculate.

(121) $1\frac{2}{36} \times 4\frac{7}{13} =$ _____

(122) $5\frac{18}{21} \times 4\frac{5}{25} =$ _____

(123) $8\frac{1}{32} \times 3\frac{13}{30} =$ _____

(124) $3\frac{7}{9} \times 8\frac{12}{22} =$ _____

(125) $6\frac{2}{4} \times 5\frac{8}{12} =$ _____

(126) $2\frac{2}{3} \times 9\frac{41}{60} =$ _____

(127) $9\frac{3}{17} \times 3\frac{41}{75} =$ _____

(128) $5\frac{15}{70} \times 8\frac{52}{100} =$ _____

(129) $7\frac{1}{2} \times 1\frac{4}{7} =$ _____

(130) $8\frac{7}{24} \times 6\frac{5}{6} =$ _____

Mixed Fractions - Multiplication

Calculate.

(131) $3\frac{18}{25} \times 9\frac{7}{9} =$ _____

(132) $6\frac{30}{40} \times 2\frac{1}{18} =$ _____

(133) $3\frac{6}{12} \times 3\frac{13}{36} =$ _____

(134) $6\frac{2}{4} \times 4\frac{1}{3} =$ _____

(135) $4\frac{13}{23} \times 4\frac{5}{20} =$ _____

(136) $2\frac{22}{32} \times 4\frac{56}{60} =$ _____

(137) $1\frac{4}{6} \times 5\frac{2}{10} =$ _____

(138) $8\frac{14}{19} \times 2\frac{1}{2} =$ _____

(139) $3\frac{36}{100} \times 6\frac{3}{14} =$ _____

(140) $3\frac{3}{21} \times 2\frac{57}{70} =$ _____

Name:
Date:
___/___/____
Time Taken:
_____ Min

Mixed Fractions - Multiplication

Calculate.

(141) $5\frac{17}{60} \times 2\frac{4}{16} =$ _____

(142) $4\frac{22}{24} \times 8\frac{1}{2} =$ _____

(143) $5\frac{13}{75} \times 9\frac{43}{100} =$ _____

(144) $9\frac{6}{8} \times 1\frac{4}{5} =$ _____

(145) $3\frac{10}{15} \times 7\frac{3}{36} =$ _____

(146) $4\frac{5}{22} \times 5\frac{1}{7} =$ _____

(147) $5\frac{2}{19} \times 5\frac{1}{17} =$ _____

(148) $1\frac{6}{8} \times 4\frac{2}{7} =$ _____

(149) $3\frac{12}{16} \times 7\frac{3}{24} =$ _____

(150) $1\frac{7}{13} \times 5\frac{3}{10} =$ _____

Mixed Fractions- Division

Calculate.

(151) $6\frac{17}{23} \div 4\frac{7}{9} =$ _____

(152) $1\frac{1}{40} \div 4\frac{4}{14} =$ _____

(153) $8\frac{54}{60} \div 4\frac{10}{20} =$ _____

(154) $5\frac{8}{21} \div 1\frac{22}{24} =$ _____

(155) $9\frac{6}{10} \div 2\frac{34}{36} =$ _____

(156) $7\frac{6}{18} \div 9\frac{2}{16} =$ _____

(157) $2\frac{14}{75} \div 1\frac{3}{11} =$ _____

(158) $2\frac{4}{5} \div 6\frac{20}{25} =$ _____

(159) $3\frac{10}{19} \div 5\frac{1}{12} =$ _____

(160) $8\frac{4}{100} \div 4\frac{1}{4} =$ _____

Mixed Fractions- Division

Calculate.

(161) $4\frac{2}{8} \div 8\frac{9}{14} =$ _____

(162) $5\frac{3}{5} \div 3\frac{4}{21} =$ _____

(163) $4\frac{4}{19} \div 8\frac{14}{16} =$ _____

(164) $8\frac{15}{32} \div 6\frac{4}{6} =$ _____

(165) $1\frac{5}{8} \div 6\frac{1}{12} =$ _____

(166) $2\frac{3}{50} \div 8\frac{4}{30} =$ _____

(167) $5\frac{2}{7} \div 2\frac{3}{36} =$ _____

(168) $6\frac{3}{17} \div 3\frac{13}{25} =$ _____

(169) $2\frac{5}{10} \div 5\frac{10}{11} =$ _____

(170) $7\frac{12}{20} \div 7\frac{1}{2} =$ _____

Mixed Fractions- Division

Calculate.

(171) $4\frac{7}{9} \div 9\frac{6}{15} =$ _____

(172) $5\frac{4}{5} \div 9\frac{24}{36} =$ _____

(173) $9\frac{3}{100} \div 8\frac{2}{11} =$ _____

(174) $7\frac{1}{18} \div 1\frac{32}{50} =$ _____

(175) $7\frac{6}{70} \div 1\frac{8}{23} =$ _____

(176) $1\frac{2}{3} \div 6\frac{3}{32} =$ _____

(177) $7\frac{4}{6} \div 7\frac{6}{16} =$ _____

(178) $1\frac{15}{17} \div 5\frac{4}{10} =$ _____

(179) $1\frac{5}{7} \div 7\frac{1}{2} =$ _____

(180) $9\frac{50}{75} \div 7\frac{7}{12} =$ _____

Name:
_____
Date:
__/__/____
Time Taken:
_____ Min

Fractions: Multiple Operations

Find the solution.

181) $(\frac{4}{5} \times \frac{1}{4}) + (\frac{3}{5} \times \frac{3}{4}) =$

_____

182) $(\frac{1}{8} + \frac{3}{8}) - (\frac{3}{8} \times \frac{5}{8}) =$

_____

183) $\frac{1}{4} + \frac{1}{6} + \frac{1}{4} + \frac{1}{6} =$

_____

184) $(\frac{4}{5} \times \frac{1}{5}) + (\frac{2}{5} \times \frac{3}{5}) =$

_____

185) $\frac{1}{3} \times \frac{2}{3} \times \frac{2}{3} =$

_____

186) $(\frac{1}{3} + \frac{1}{3}) \div \frac{2}{3} =$

_____

187) $(\frac{7}{8} + \frac{5}{8}) \times (\frac{7}{8} + \frac{3}{8}) =$

_____

188) $\frac{1}{5} + \frac{5}{6} + \frac{2}{5} + \frac{1}{6} =$

_____

189) $\frac{1}{6} + \frac{2}{3} - \frac{5}{6} =$

_____

190) $\frac{1}{4} \times \frac{5}{8} + \frac{3}{8} =$

_____

Fractions: Multiple Operations

Find the solution.

(191) $\frac{1}{4} + \frac{1}{6} + 8 =$

_____

(192) $(\frac{3}{8} + \frac{3}{8}) \times (\frac{3}{8} + \frac{1}{8}) =$

_____

(193) $(\frac{1}{4} + \frac{3}{4}) \div \frac{1}{4} =$

_____

(194) $\frac{3}{5} + \frac{1}{4} + \frac{2}{5} + \frac{1}{4} =$

_____

(195) $(\frac{5}{6} + \frac{1}{6}) - (\frac{1}{3} \times \frac{2}{3}) =$

_____

(196) $\frac{5}{6} + \frac{2}{5} - \frac{1}{6} =$

_____

(197) $(\frac{3}{5} + \frac{4}{5}) \times (\frac{3}{5} + \frac{1}{5}) =$

_____

(198) $\frac{2}{3} + \frac{1}{3} + \frac{2}{3} =$

_____

(199) $\frac{5}{8} + \frac{3}{8} + \frac{3}{8} =$

_____

(200) $\frac{1}{6} + \frac{1}{6} + \frac{5}{6} =$

_____

Name:
_____
Date:
__/__/____
Time Taken:
_____ Min

Fractions: Multiple Operations

Find the solution.

(201) $\frac{1}{3} + \frac{2}{3} - \frac{2}{3} =$

_____

(202) $\left(\frac{1}{8} + \frac{3}{8}\right) \div \frac{7}{8} =$

_____

(203) $\frac{2}{3} + \frac{2}{3} + \frac{2}{3} =$

_____

(204) $\left(\frac{2}{3} + \frac{2}{3}\right) - \left(\frac{3}{4} \times \frac{1}{4}\right) =$

_____

(205) $\left(\frac{1}{6} + \frac{5}{6}\right) \times \left(\frac{1}{6} + \frac{1}{6}\right) =$

_____

(206) $\frac{2}{3} + \frac{1}{6} + 3 =$

_____

(207) $\left(\frac{7}{8} + \frac{5}{8}\right) \times \left(\frac{5}{8} + \frac{1}{8}\right) =$

_____

(208) $\frac{1}{3} \times \frac{3}{4} + \frac{3}{4} =$

_____

(209) $\frac{1}{6} \times \frac{1}{6} \times \frac{1}{6} =$

_____

(210) $\left(\frac{5}{8} + \frac{7}{8}\right) \div \frac{1}{8} =$

_____

Name:
_____
Date:
__/__/____
Time Taken:
_____ Min

Fractions: Multiple Operations

Find the solution.

(211) $\frac{1}{3} + \frac{4}{5} + 1 =$

_____

(212) $\frac{5}{8} \times \frac{7}{8} + \frac{3}{8} =$

_____

(213) $\frac{1}{6} \times \frac{3}{4} + \frac{3}{4} =$

_____

(214) $\frac{3}{4} + \frac{1}{6} + \frac{3}{4} + \frac{1}{6} =$

_____

(215) $\left(\frac{2}{3} \times \frac{1}{8}\right) + \left(\frac{1}{3} \times \frac{3}{8}\right) =$

_____

(216) $\frac{4}{5} \times \frac{1}{6} + \frac{1}{6} =$

_____

(217) $\left(\frac{3}{4} + \frac{3}{4}\right) \div \frac{1}{4} =$

_____

(218) $\frac{1}{5} \times \frac{4}{5} \times \frac{4}{5} =$

_____

(219) $\frac{1}{4} + \frac{5}{6} + 4 =$

_____

(220) $\left(\frac{1}{3} + \frac{1}{3}\right) \times \left(\frac{1}{3} + \frac{2}{3}\right) =$

_____

Fractions: Multiple Operations

Find the solution.

(221) $\left(\dfrac{2}{3} + \dfrac{1}{3}\right) - \left(\dfrac{1}{4} \times \dfrac{3}{4}\right) =$ _____

(222) $\dfrac{2}{5} + \dfrac{5}{8} + 5 =$ _____

(223) $\dfrac{1}{6} \times \dfrac{1}{6} \times \dfrac{1}{6} =$ _____

(224) $\dfrac{5}{8} + \dfrac{3}{8} - \dfrac{1}{8} =$ _____

(225) $\dfrac{1}{3} + \dfrac{1}{3} + \dfrac{2}{3} =$ _____

(226) $\dfrac{5}{6} + \dfrac{1}{3} + 7 =$ _____

(227) $\left(\dfrac{4}{5} + \dfrac{2}{5}\right) \times \left(\dfrac{1}{5} + \dfrac{2}{5}\right) =$ _____

(228) $\dfrac{1}{4} + \dfrac{3}{4} + \dfrac{3}{4} =$ _____

(229) $\dfrac{1}{8} + \dfrac{5}{8} + \dfrac{1}{8} =$ _____

(230) $\dfrac{3}{5} + \dfrac{5}{8} + \dfrac{3}{5} + \dfrac{7}{8} =$ _____

Fractions: Multiple Operations

Find the solution.

(231) $\frac{4}{5} \times \frac{1}{4} + \frac{1}{4} =$

_____

(232) $\frac{2}{3} \times \frac{1}{3} \times \frac{1}{3} =$

_____

(233) $(\frac{1}{4} + \frac{1}{4}) - (\frac{3}{8} \times \frac{3}{8}) =$

_____

(234) $\frac{1}{6} + \frac{4}{5} - \frac{1}{6} =$

_____

(235) $\frac{1}{4} + \frac{5}{8} + \frac{1}{4} + \frac{1}{8} =$

_____

(236) $\frac{2}{5} + \frac{4}{5} + \frac{3}{5} + \frac{1}{5} =$

_____

(237) $\frac{1}{6} \times \frac{2}{3} + \frac{2}{3} =$

_____

(238) $\frac{1}{3} + \frac{4}{5} + 5 =$

_____

(239) $\frac{3}{4} \times \frac{3}{5} + \frac{4}{5} =$

_____

(240) $(\frac{5}{6} + \frac{1}{6}) \times (\frac{1}{6} + \frac{1}{6}) =$

_____

Fractions: Multiple Operations

Find the solution.

(241) $\frac{1}{6} + \frac{1}{6} + \frac{1}{6} =$

_____

(242) $\frac{2}{3} \times \frac{1}{6} + \frac{1}{6} =$

_____

(243) $\frac{2}{3} \times \frac{2}{3} \times \frac{2}{3} =$

_____

(244) $\frac{3}{5} + \frac{5}{8} - \frac{2}{5} =$

_____

(245) $\left(\frac{1}{4} + \frac{1}{4}\right) \times \left(\frac{1}{4} + \frac{1}{4}\right) =$

_____

(246) $\left(\frac{1}{5} + \frac{1}{5}\right) \times \left(\frac{4}{5} + \frac{4}{5}\right) =$

_____

(247) $\frac{5}{8} + \frac{3}{8} + 5 =$

_____

(248) $\left(\frac{1}{4} \times \frac{2}{3}\right) + \left(\frac{1}{4} \times \frac{2}{3}\right) =$

_____

(249) $\left(\frac{1}{4} + \frac{1}{4}\right) - \left(\frac{3}{5} \times \frac{3}{5}\right) =$

_____

(250) $\left(\frac{2}{5} + \frac{3}{5}\right) \times \left(\frac{1}{5} + \frac{1}{5}\right) =$

_____

Fractions: Multiple Operations

Find the solution.

(251) $\dfrac{1}{8} \times \dfrac{1}{5} + \dfrac{2}{5} =$

_____

(252) $\left(\dfrac{3}{4} + \dfrac{1}{4}\right) \times \left(\dfrac{1}{4} + \dfrac{3}{4}\right) =$

_____

(253) $\dfrac{1}{4} \times \dfrac{1}{5} + \dfrac{1}{5} =$

_____

(254) $\left(\dfrac{1}{4} + \dfrac{1}{4}\right) - \left(\dfrac{2}{5} \times \dfrac{2}{5}\right) =$

_____

(255) $\left(\dfrac{5}{8} + \dfrac{1}{8}\right) - \left(\dfrac{5}{8} \times \dfrac{5}{8}\right) =$

_____

(256) $\dfrac{2}{3} \times \dfrac{1}{3} \times \dfrac{1}{3} =$

_____

(257) $\dfrac{1}{6} + \dfrac{1}{6} + \dfrac{5}{6} =$

_____

(258) $\dfrac{7}{8} + \dfrac{5}{8} - \dfrac{1}{8} =$

_____

(259) $\dfrac{2}{3} + \dfrac{1}{3} + \dfrac{1}{3} =$

_____

(260) $\dfrac{3}{8} \times \dfrac{7}{8} \times \dfrac{5}{8} =$

_____

Simplifying Fractions

(261) $\frac{12}{72}$ = _____

(262) $\frac{400}{60}$ = _____

(263) $\frac{45}{15}$ = _____

(264) $\frac{486}{60}$ = _____

(265) $\frac{18}{36}$ = _____

(266) $\frac{210}{35}$ = _____

(267) $\frac{25}{60}$ = _____

(268) $\frac{56}{112}$ = _____

(269) $\frac{336}{42}$ = _____

(270) $\frac{45}{135}$ = _____

(271) $\frac{255}{30}$ = _____

(272) $\frac{42}{18}$ = _____

(273) $\frac{384}{48}$ = _____

(274) $\frac{120}{16}$ = _____

(275) $\frac{280}{40}$ = _____

(276) $\frac{16}{48}$ = _____

(277) $\frac{672}{84}$ = _____

(278) $\frac{36}{9}$ = _____

(279) $\frac{532}{56}$ = _____

(280) $\frac{270}{90}$ = _____

Simplifying Fractions

281) $\dfrac{240}{30}$ = _____

282) $\dfrac{12}{32}$ = _____

283) $\dfrac{8}{56}$ = _____

284) $\dfrac{40}{8}$ = _____

285) $\dfrac{49}{105}$ = _____

286) $\dfrac{480}{96}$ = _____

287) $\dfrac{192}{48}$ = _____

288) $\dfrac{736}{112}$ = _____

289) $\dfrac{60}{30}$ = _____

290) $\dfrac{18}{45}$ = _____

291) $\dfrac{306}{72}$ = _____

292) $\dfrac{14}{21}$ = _____

293) $\dfrac{28}{40}$ = _____

294) $\dfrac{18}{48}$ = _____

295) $\dfrac{6}{24}$ = _____

296) $\dfrac{176}{24}$ = _____

297) $\dfrac{40}{112}$ = _____

298) $\dfrac{36}{54}$ = _____

299) $\dfrac{9}{36}$ = _____

300) $\dfrac{160}{20}$ = _____

Simplifying Fractions

(301) $\frac{76}{12}$ = _____

(302) $\frac{60}{10}$ = _____

(303) $\frac{420}{60}$ = _____

(304) $\frac{32}{40}$ = _____

(305) $\frac{45}{15}$ = _____

(306) $\frac{352}{48}$ = _____

(307) $\frac{305}{70}$ = _____

(308) $\frac{576}{96}$ = _____

(309) $\frac{18}{36}$ = _____

(310) $\frac{352}{64}$ = _____

(311) $\frac{16}{48}$ = _____

(312) $\frac{224}{24}$ = _____

(313) $\frac{114}{28}$ = _____

(314) $\frac{26}{8}$ = _____

(315) $\frac{56}{70}$ = _____

(316) $\frac{88}{120}$ = _____

(317) $\frac{505}{60}$ = _____

(318) $\frac{174}{30}$ = _____

(319) $\frac{384}{48}$ = _____

(320) $\frac{6}{10}$ = _____

Simplifying Fractions

(321) $\dfrac{480}{60}$ = _____

(322) $\dfrac{576}{96}$ = _____

(323) $\dfrac{80}{40}$ = _____

(324) $\dfrac{144}{24}$ = _____

(325) $\dfrac{16}{20}$ = _____

(326) $\dfrac{2}{12}$ = _____

(327) $\dfrac{210}{70}$ = _____

(328) $\dfrac{140}{21}$ = _____

(329) $\dfrac{14}{16}$ = _____

(330) $\dfrac{6}{30}$ = _____

(331) $\dfrac{9}{18}$ = _____

(332) $\dfrac{40}{10}$ = _____

(333) $\dfrac{3}{9}$ = _____

(334) $\dfrac{720}{90}$ = _____

(335) $\dfrac{108}{36}$ = _____

(336) $\dfrac{242}{28}$ = _____

(337) $\dfrac{288}{48}$ = _____

(338) $\dfrac{30}{60}$ = _____

(339) $\dfrac{224}{32}$ = _____

(340) $\dfrac{20}{25}$ = _____

Simplifying Fractions

(341) $\frac{424}{48}$ = _____

(342) $\frac{49}{56}$ = _____

(343) $\frac{525}{105}$ = _____

(344) $\frac{90}{30}$ = _____

(345) $\frac{66}{9}$ = _____

(346) $\frac{258}{48}$ = _____

(347) $\frac{889}{105}$ = _____

(348) $\frac{1026}{108}$ = _____

(349) $\frac{8}{20}$ = _____

(350) $\frac{927}{126}$ = _____

(351) $\frac{180}{36}$ = _____

(352) $\frac{60}{12}$ = _____

(353) $\frac{350}{50}$ = _____

(354) $\frac{6}{18}$ = _____

(355) $\frac{20}{40}$ = _____

(356) $\frac{56}{84}$ = _____

(357) $\frac{14}{28}$ = _____

(358) $\frac{672}{84}$ = _____

(359) $\frac{162}{72}$ = _____

(360) $\frac{45}{54}$ = _____

Fast Math Success Workbook Grade 6-7

Simplifying Fractions

(361) $\dfrac{90}{15}$ = _____

(362) $\dfrac{6}{20}$ = _____

(363) $\dfrac{9}{27}$ = _____

(364) $\dfrac{10}{20}$ = _____

(365) $\dfrac{448}{56}$ = _____

(366) $\dfrac{360}{45}$ = _____

(367) $\dfrac{288}{96}$ = _____

(368) $\dfrac{64}{120}$ = _____

(369) $\dfrac{10}{30}$ = _____

(370) $\dfrac{16}{32}$ = _____

(371) $\dfrac{12}{16}$ = _____

(372) $\dfrac{210}{35}$ = _____

(373) $\dfrac{168}{84}$ = _____

(374) $\dfrac{64}{32}$ = _____

(375) $\dfrac{238}{56}$ = _____

(376) $\dfrac{351}{42}$ = _____

(377) $\dfrac{160}{50}$ = _____

(378) $\dfrac{984}{120}$ = _____

(379) $\dfrac{4}{12}$ = _____

(380) $\dfrac{12}{18}$ = _____

Simplifying Fractions

381) $\frac{252}{84}$ = _____

382) $\frac{54}{90}$ = _____

383) $\frac{189}{28}$ = _____

384) $\frac{63}{21}$ = _____

385) $\frac{945}{135}$ = _____

386) $\frac{80}{25}$ = _____

387) $\frac{35}{42}$ = _____

388) $\frac{612}{72}$ = _____

389) $\frac{27}{72}$ = _____

390) $\frac{2}{8}$ = _____

391) $\frac{35}{98}$ = _____

392) $\frac{40}{48}$ = _____

393) $\frac{399}{56}$ = _____

394) $\frac{252}{45}$ = _____

395) $\frac{21}{9}$ = _____

396) $\frac{261}{36}$ = _____

397) $\frac{420}{60}$ = _____

398) $\frac{155}{30}$ = _____

399) $\frac{70}{75}$ = _____

400) $\frac{95}{20}$ = _____

Percent

Find the percentage of given numbers and percent values.

(401) 15% of 869 = [   ]

(402) 20% of 886 = [   ]

(403) 5% of 30 = [   ]

(404) 10% of [   ] = 31.2

(405) [   ] of 526 = 131.5

(406) 8% of 525 = [   ]

(407) [   ] of 965 = 19.3

(408) [   ] of 909 = 136.35

(409) 2% of [   ] = 16.72

(410) 20% of 684 = [   ]

(411) [   ] of 948 = 237

(412) 8% of [   ] = 22.08

(413) 5% of 336 = [   ]

(414) [   ] of 94 = 9.4

(415) 25% of 385 = [   ]

(416) [   ] of 178 = 17.8

(417) 2% of [   ] = 9.94

(418) [   ] of 183 = 14.64

(419) [   ] of 355 = 53.25

(420) [   ] of 687 = 34.35

Percent

Find the percentage of given numbers and percent values.

(421) [ ] of 710 = 106.5

(422) [ ] of 663 = 66.3

(423) 8% of [ ] = 41.6

(424) 5% of 411 = [ ]

(425) 15% of [ ] = 22.65

(426) [ ] of 567 = 141.75

(427) [ ] of 334 = 6.68

(428) 20% of [ ] = 29.4

(429) 2% of 564 = [ ]

(430) 10% of 696 = [ ]

(431) [ ] of 981 = 147.15

(432) 25% of [ ] = 184

(433) 8% of 860 = [ ]

(434) 20% of [ ] = 103.2

(435) 5% of [ ] = 41.8

(436) 15% of 196 = [ ]

(437) 8% of [ ] = 59.44

(438) 25% of 502 = [ ]

(439) 5% of 398 = [ ]

(440) 10% of 470 = [ ]

Percent

Find the percentage of given numbers and percent values.

(441) [   ] of 747 = 37.35

(442) 15% of 743 = [   ]

(443) 2% of 256 = [   ]

(444) 5% of [   ] = 17.3

(445) 15% of [   ] = 132.45

(446) 10% of [   ] = 43.5

(447) 25% of [   ] = 29.5

(448) [   ] of 290 = 58

(449) 8% of [   ] = 50

(450) 25% of 140 = [   ]

(451) [   ] of 686 = 54.88

(452) 15% of 373 = [   ]

(453) 5% of 993 = [   ]

(454) 20% of 622 = [   ]

(455) 2% of [   ] = 7.08

(456) 10% of [   ] = 55.3

(457) 8% of [   ] = 54.4

(458) [   ] of 971 = 242.75

(459) [   ] of 272 = 27.2

(460) 15% of 433 = [   ]

Name: _ _ _ _ _ _ _ _ _

Date: _ _ / _ _ / _ _ _

Time Taken: _ _ _ _ Min

Percent

Find the percentage of given numbers and percent values.

(461) 5% of 239 = ⬚

(462) 20% of ⬚ = 28.6

(463) ⬚ of 672 = 168

(464) 10% of ⬚ = 65.7

(465) 25% of ⬚ = 61.5

(466) 15% of ⬚ = 48.6

(467) 20% of 20 = ⬚

(468) 8% of ⬚ = 60.8

(469) 2% of ⬚ = 6.52

(470) ⬚ of 134 = 6.7

(471) 2% of 432 = ⬚

(472) 10% of 302 = ⬚

(473) ⬚ of 877 = 43.85

(474) 25% of ⬚ = 135

(475) ⬚ of 349 = 69.8

(476) 8% of ⬚ = 45.76

(477) 15% of ⬚ = 91.05

(478) 15% of 437 = ⬚

(479) 2% of ⬚ = 1.68

(480) 10% of 703 = ⬚

Name:
_____

Date:
___/___/____

Time Taken:
_____ Min

Percent

Find the percentage of given numbers and percent values.

(481) 8% of [ ] = 35.84

(482) 10% of 105 = [ ]

(483) [ ] of 300 = 75

(484) 2% of 144 = [ ]

(485) 10% of 632 = [ ]

(486) [ ] of 537 = 80.55

(487) 2% of 974 = [ ]

(488) 20% of 991 = [ ]

(489) 5% of 246 = [ ]

(490) 8% of 592 = [ ]

(491) 25% of [ ] = 41.5

(492) [ ] of 276 = 41.4

(493) 8% of 464 = [ ]

(494) [ ] of 469 = 23.45

(495) [ ] of 29 = 5.8

(496) 25% of [ ] = 52.25

(497) 10% of [ ] = 59.7

(498) [ ] of 108 = 2.16

(499) 25% of 957 = [ ]

(500) 10% of 419 = [ ]

Percent and Decimals

Convert Percent to Decimal.

(501) 85 % = _____

(502) 29 % = _____

(503) 52 % = _____

(504) 16 % = _____

(505) 25 % = _____

(506) 56 % = _____

(507) 53 % = _____

(508) 44 % = _____

(509) 84 % = _____

(510) 7 % = _____

(511) 62 % = _____

(512) 32 % = _____

(513) 82 % = _____

(514) 35 % = _____

(515) 34 % = _____

(516) 47 % = _____

(517) 75 % = _____

(518) 54 % = _____

(519) 72 % = _____

(520) 93 % = _____

Percent and Decimals

Convert Percent to Decimal.

(521) 21 % = _____

(522) 11 % = _____

(523) 57 % = _____

(524) 97 % = _____

(525) 30 % = _____

(526) 82 % = _____

(527) 67 % = _____

(528) 98 % = _____

(529) 29 % = _____

(530) 39 % = _____

(531) 10 % = _____

(532) 61 % = _____

(533) 13 % = _____

(534) 68 % = _____

(535) 38 % = _____

(536) 9 % = _____

(537) 76 % = _____

(538) 79 % = _____

(539) 75 % = _____

(540) 37 % = _____

FAST
MATH
SUCCESS

Name:
_____

Date:
__/__/____

Time Taken:
____ Min

Percent and Decimals

Convert Percent to Decimal.

(541) 25 % = _____

(542) 73 % = _____

(543) 70 % = _____

(544) 80 % = _____

(545) 46 % = _____

(546) 66 % = _____

(547) 78 % = _____

(548) 93 % = _____

(549) 11 % = _____

(550) 18 % = _____

(551) 42 % = _____

(552) 69 % = _____

(553) 33 % = _____

(554) 72 % = _____

(555) 92 % = _____

(556) 32 % = _____

(557) 79 % = _____

(558) 88 % = _____

(559) 54 % = _____

(560) 59 % = _____

Percent and Decimals
Convert Percent to Decimal.

(561) 79 % = _____

(562) 90 % = _____

(563) 50 % = _____

(564) 36 % = _____

(565) 6 % = _____

(566) 82 % = _____

(567) 66 % = _____

(568) 93 % = _____

(569) 95 % = _____

(570) 20 % = _____

(571) 21 % = _____

(572) 77 % = _____

(573) 100 % = _____

(574) 52 % = _____

(575) 69 % = _____

(576) 71 % = _____

(577) 43 % = _____

(578) 44 % = _____

(579) 98 % = _____

(580) 88 % = _____

Percent and Decimals
Convert Decimal to Percent.

(581) 0.2 = _____

(582) 0.54 = _____

(583) 0.22 = _____

(584) 0.91 = _____

(585) 0.64 = _____

(586) 0.96 = _____

(587) 0.39 = _____

(588) 0.95 = _____

(589) 0.49 = _____

(590) 0.44 = _____

(591) 0.5 = _____

(592) 0.84 = _____

(593) 0.47 = _____

(594) 0.41 = _____

(595) 0.85 = _____

(596) 0.12 = _____

(597) 0.01 = _____

(598) 0.17 = _____

(599) 0.99 = _____

(600) 0.21 = _____

Percent and Decimals

Convert Decimal to Percent.

(601) 0.65 = _____

(602) 0.7 = _____

(603) 0.33 = _____

(604) 0.2 = _____

(605) 0.24 = _____

(606) 0.61 = _____

(607) 0.36 = _____

(608) 0.44 = _____

(609) 0.56 = _____

(610) 0.72 = _____

(611) 0.59 = _____

(612) 0.29 = _____

(613) 0.46 = _____

(614) 0.42 = _____

(615) 0.87 = _____

(616) 0.28 = _____

(617) 0.99 = _____

(618) 0.47 = _____

(619) 0.81 = _____

(620) 0.78 = _____

Percent and Decimals

Convert Decimal to Percent.

(621) 0.28 = _____

(622) 0.16 = _____

(623) 0.53 = _____

(624) 0.69 = _____

(625) 0.49 = _____

(626) 0.27 = _____

(627) 0.03 = _____

(628) 0.6 = _____

(629) 0.87 = _____

(630) 0.04 = _____

(631) 0.56 = _____

(632) 0.02 = _____

(633) 0.94 = _____

(634) 0.63 = _____

(635) 0.19 = _____

(636) 0.71 = _____

(637) 0.88 = _____

(638) 1 = _____

(639) 0.95 = _____

(640) 0.12 = _____

Percent and Decimals

Convert Decimal to Percent.

(641) 0.88 = _____

(642) 0.48 = _____

(643) 0.92 = _____

(644) 0.99 = _____

(645) 0.84 = _____

(646) 0.03 = _____

(647) 0.7 = _____

(648) 0.18 = _____

(649) 0.43 = _____

(650) 0.38 = _____

(651) 0.55 = _____

(652) 0.8 = _____

(653) 0.52 = _____

(654) 0.75 = _____

(655) 0.34 = _____

(656) 0.33 = _____

(657) 0.85 = _____

(658) 0.69 = _____

(659) 0.86 = _____

(660) 0.56 = _____

Percent - Advanced

Calculate the given percent of each value.

(661) [ ] of 78 = 1.326

(662) [ ] of 93 = 3.72

(663) [ ] of 5 = 0.185

(664) 9.0% of [ ] = 56.97

(665) 0.8% of [ ] = 2.216

(666) 5.7% of 516 = [ ]

(667) [ ] of 37 = 0.148

(668) 2.9% of 898 = [ ]

(669) 0.5% of [ ] = 3.035

(670) 9.1% of [ ] = 0.546

(671) [ ] of 92 = 4.6

(672) 8.2% of 4 = [ ]

(673) [ ] of 7 = 0.028

(674) [ ] of 168 = 13.104

(675) 8.5% of [ ] = 0.765

(676) 0.5% of 268 = [ ]

(677) [ ] of 548 = 1.096

(678) 0.5% of 87 = [ ]

(679) 6.8% of 534 = [ ]

(680) 0.9% of 19 = [ ]

Percent - Advanced

Calculate the given percent of each value.

(681) 9.0% of 7 = [＿＿＿]

(682) 5.7% of [＿＿] = 1.425

(683) 3.3% of 983 = [＿＿＿]

(684) [＿＿] of 32 = 1.184

(685) 3.0% of 761 = [＿＿＿]

(686) 0.1% of [＿＿] = 0.613

(687) 2.8% of [＿＿] = 0.168

(688) 4.3% of [＿＿] = 40.893

(689) [＿＿] of 144 = 10.512

(690) 1.5% of [＿＿] = 0.09

(691) [＿＿] of 16 = 0.032

(692) 0.6% of 7 = [＿＿＿]

(693) [＿＿] of 5 = 0.395

(694) [＿＿] of 369 = 1.845

(695) 3.9% of [＿＿] = 1.56

(696) 2.2% of 818 = [＿＿＿]

(697) 0.6% of 74 = [＿＿＿]

(698) 4.8% of 232 = [＿＿＿]

(699) 9.6% of [＿＿] = 92.16

(700) [＿＿] of 3 = 0.102

Percent - Advanced

Calculate the given percent of each value.

(701) 2.7% of 646 = [　　　]

(702) 7.9% of [　　　] = 19.829

(703) [　　　] of 41 = 2.501

(704) 6.2% of 2 = [　　　]

(705) [　　　] of 98 = 2.94

(706) 0.1% of [　　　] = 0.977

(707) 0.7% of [　　　] = 0.441

(708) 5.7% of 10 = [　　　]

(709) [　　　] of 5 = 0.035

(710) [　　　] of 6 = 0.528

(711) 1.2% of [　　　] = 0.708

(712) 2.2% of 2 = [　　　]

(713) 1.0% of 765 = [　　　]

(714) 9.0% of [　　　] = 23.67

(715) [　　　] of 825 = 28.875

(716) 7.3% of 627 = [　　　]

(717) [　　　] of 9 = 0.045

(718) [　　　] of 4 = 0.024

(719) 1.7% of 5 = [　　　]

(720) 0.5% of 67 = [　　　]

Percent - Advanced

Calculate the given percent of each value.

(721) 3.2% of 3 = [ ]

(722) 1.5% of 5 = [ ]

(723) [ ] of 5 = 0.215

(724) 0.4% of [ ] = 0.044

(725) 0.9% of 74 = [ ]

(726) 8.0% of [ ] = 3.6

(727) [ ] of 803 = 4.015

(728) 3.7% of 509 = [ ]

(729) 0.1% of 7 = [ ]

(730) [ ] of 717 = 1.434

(731) 9.6% of 340 = [ ]

(732) 0.6% of [ ] = 0.024

(733) 0.6% of [ ] = 3.756

(734) 0.5% of 1 = [ ]

(735) [ ] of 4 = 0.008

(736) 0.8% of [ ] = 0.048

(737) 2.7% of [ ] = 21.168

(738) 9.6% of [ ] = 59.136

(739) 0.9% of [ ] = 4.554

(740) [ ] of 50 = 1.65

Name:

Date:
___/___/____

Time Taken:
____ Min

Percent - Advanced

Calculate the given percent of each value.

(741) 8.8% of 9 = [   ]

(742) 8.7% of [   ] = 0.174

(743) [   ] of 4 = 0.18

(744) [   ] of 2 = 0.062

(745) 8.4% of [   ] = 72.072

(746) 0.5% of [   ] = 0.035

(747) 5.3% of [   ] = 2.597

(748) 2.8% of [   ] = 2.744

(749) 2.3% of [   ] = 2.369

(750) 0.5% of 905 = [   ]

(751) [   ] of 580 = 37.7

(752) 7.7% of [   ] = 12.859

(753) [   ] of 327 = 0.654

(754) 0.8% of 67 = [   ]

(755) 5.2% of [   ] = 24.544

(756) [   ] of 629 = 8.806

(757) 3.6% of 55 = [   ]

(758) 3.4% of 68 = [   ]

(759) [   ] of 3 = 0.267

(760) 0.7% of [   ] = 0.014

Percent - Advanced

Calculate the given percent of each value.

(761) 3.6% of [＿＿] = 17.676

(762) 0.2% of 350 = [＿＿]

(763) [＿＿] of 3 = 0.21

(764) 0.5% of [＿＿] = 0.33

(765) [＿＿] of 3 = 0.177

(766) 1.2% of 8 = [＿＿]

(767) [＿＿] of 965 = 0.965

(768) 0.4% of 503 = [＿＿]

(769) [＿＿] of 61 = 0.183

(770) 1.4% of 87 = [＿＿]

(771) 1.4% of [＿＿] = 9.786

(772) [＿＿] of 8 = 0.016

(773) [＿＿] of 490 = 18.62

(774) 0.1% of [＿＿] = 0.966

(775) [＿＿] of 495 = 7.92

(776) 0.8% of 372 = [＿＿]

(777) [＿＿] of 36 = 2.268

(778) [＿＿] of 10 = 0.64

(779) 0.3% of 2 = [＿＿]

(780) 2.5% of 41 = [＿＿]

Percent - Advanced

Calculate the given percent of each value.

(781) [____] of 85 = 0.51

(782) 5.7% of 755 = [____]

(783) [____] of 850 = 1.7

(784) [____] of 943 = 50.922

(785) [____] of 98 = 2.45

(786) 1.4% of [____] = 1.106

(787) 1.2% of [____] = 0.084

(788) [____] of 8 = 0.048

(789) 0.8% of [____] = 0.016

(790) 9.4% of 140 = [____]

(791) 0.3% of [____] = 1.425

(792) [____] of 1 = 0.029

(793) 8.0% of 95 = [____]

(794) 0.1% of [____] = 0.002

(795) [____] of 82 = 0.246

(796) 3.5% of 49 = [____]

(797) 4.4% of 6 = [____]

(798) 0.3% of 67 = [____]

(799) 7.7% of [____] = 63.679

(800) 7.7% of [____] = 0.385

Percent - Advanced

Calculate the given percent of each value.

(801) 0.5% of ⬚ = 0.035

(802) 8.2% of ⬚ = 0.328

(803) 0.2% of 2 = ⬚

(804) ⬚ of 59 = 2.714

(805) 6.9% of ⬚ = 67.206

(806) ⬚ of 83 = 8.134

(807) 0.7% of ⬚ = 0.21

(808) 8.2% of 8 = ⬚

(809) 0.2% of ⬚ = 0.104

(810) 4.2% of ⬚ = 3.738

(811) 5.9% of ⬚ = 0.177

(812) ⬚ of 6 = 0.006

(813) 1.2% of ⬚ = 0.228

(814) 6.8% of ⬚ = 2.72

(815) 6.3% of ⬚ = 0.126

(816) 0.5% of 15 = ⬚

(817) 5.9% of 346 = ⬚

(818) 0.9% of ⬚ = 8.388

(819) 8.9% of ⬚ = 4.183

(820) 2.8% of 13 = ⬚

Ratio Conversions

Provide the conversions for each ratio (Part to Part).

(821)

	Ratio	Fraction	Percent	Decimal
a.	2:2			
b.	4:6			
c.	4:5			
d.	1:8			
e.	3:6			
f.	2:8			
g.	2:3			
h.	1:9			
i.	3:9			
j.	5:6			
k.	5:7			
l.	1:3			
m.	7:9			

Ratio Conversions

Provide the conversions for each ratio (Part to Part).

(822)

	Ratio	Fraction	Percent	Decimal
a.	2:10			
b.	6:7			
c.	7:8			
d.	2:3			
e.	5:5			
f.	1:6			
g.	1:7			
h.	3:7			
i.	4:8			
j.	9:10			
k.	2:8			
l.	6:8			
m.	1:2			

Ratio Conversions

Provide the conversions for each ratio (Part to Part).

(823)

	Ratio	Fraction	Percent	Decimal
a.	2:4			
b.				0.5
c.				0.667
d.			20%	
e.				0.8
f.			75%	
g.			100%	
h.		2/7		
i.				0.333
j.		5/7		
k.	4:9			
l.	3:8			
m.				0.6

Ratio Conversions

Provide the conversions for each ratio (Part to Part).

(824)

	Ratio	Fraction	Percent	Decimal
a.		2/4		
b.	1:1			
c.	1:7			
d.			66.7%	
e.				0.6
f.		1/2		
g.	2:9			
h.			57.1%	
i.			33.3%	
j.	5:7			
k.	3:4			
l.				0.125
m.	2:6			

Cartesian Coordinates

Fill in as indicated.

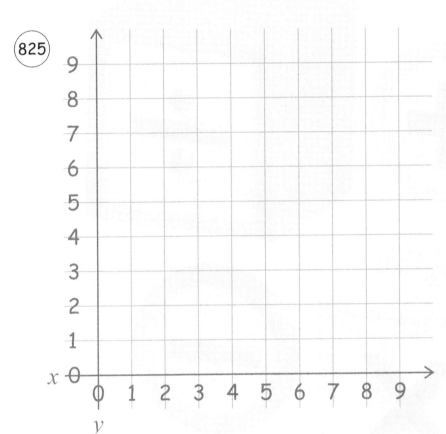

(825)

A = (9, 8)   B = (4, 6)

C = (4, 7)   D = (4, 8)

E = (2, 4)   F = (9, 4)

G = (0, 6)   H = (8, 0)

I = (9, 7)   J = (0, 5)

Cartesian Coordinates

Fill in as indicated.

(826)

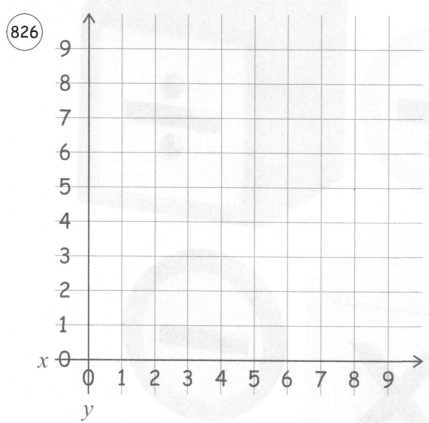

A = (9, 0)   B = (2, 5)

C = (2, 7)   D = (1, 7)

E = (2, 3)   F = (2, 4)

G = (6, 6)   H = (4, 3)

I = (3, 3)   J = (0, 0)

Name: _____

Date: __/__/__

Time Taken: ____ Min

Cartesian Coordinates

Fill in as indicated.

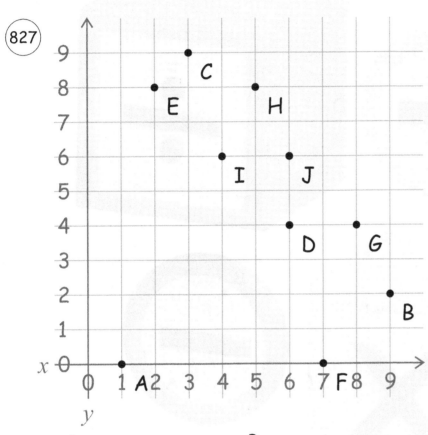

A = _____   B = _____

C = _____   D = _____

E = _____   F = _____

G = _____   H = _____

I = _____   J = _____

Cartesian Coordinates

Fill in as indicated.

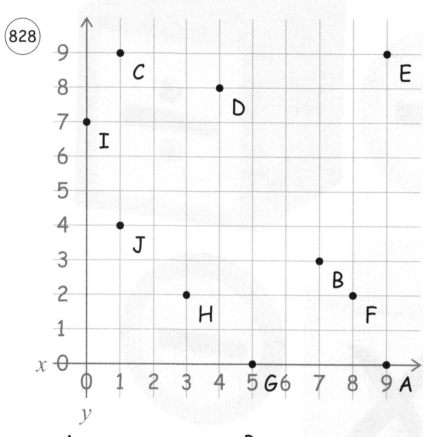

A = _____   B = _____

C = _____   D = _____

E = _____   F = _____

G = _____   H = _____

I = _____   J = _____

Cartesian Coordinates

Fill in as indicated.

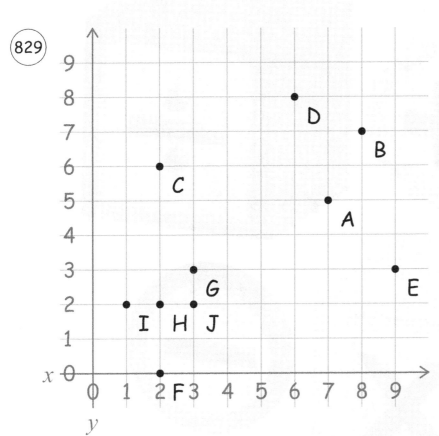

A = _____   B = _____

C = _____   D = _____

E = _____   F = _____

G = _____   H = _____

I = _____   J = _____

Cartesian Coordinates With Four Quadrants

Fill in as indicated.

(830)

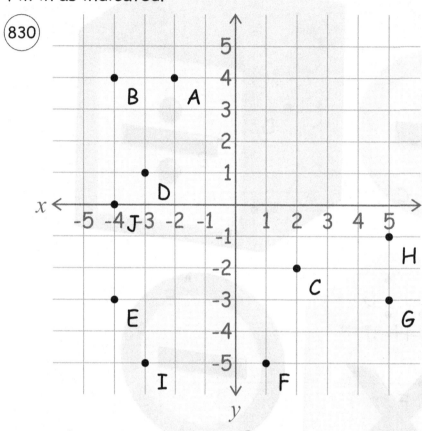

A = _____   B = _____

C = _____   D = _____

E = _____   F = _____

G = _____   H = _____

I = _____   J = _____

Name:
_____
Date:
__/__/____
Time Taken:
_____ Min

Cartesian Coordinates With Four Quadrants

Fill in as indicated.

(831)

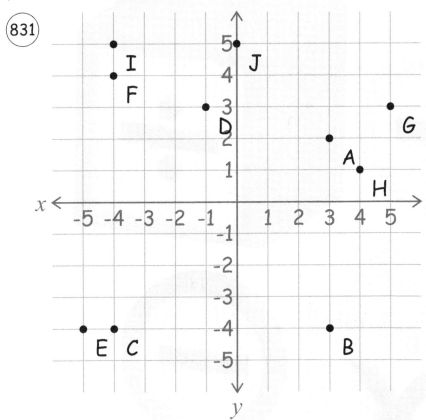

A = _____   B = _____

C = _____   D = _____

E = _____   F = _____

G = _____   H = _____

I = _____   J = _____

Cartesian Coordinates With Four Quadrants

Fill in as indicated.

(832)

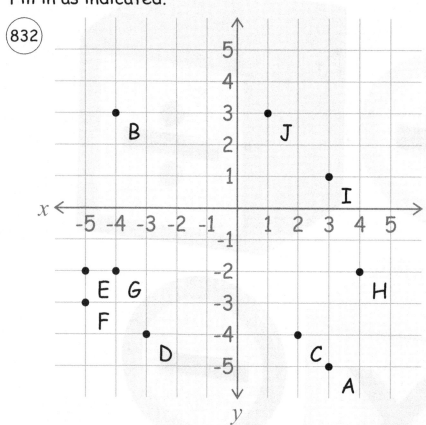

A = _____   B = _____

C = _____   D = _____

E = _____   F = _____

G = _____   H = _____

I = _____   J = _____

Cartesian Coordinates With Four Quadrants

Fill in as indicated.

(833)

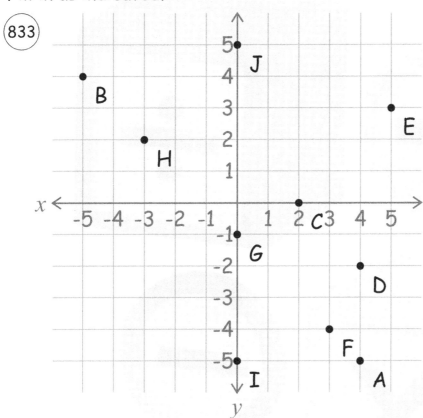

A = _____   B = _____

C = _____   D = _____

E = _____   F = _____

G = _____   H = _____

I = _____   J = _____

Cartesian Coordinates With Four Quadrants

Fill in as indicated.

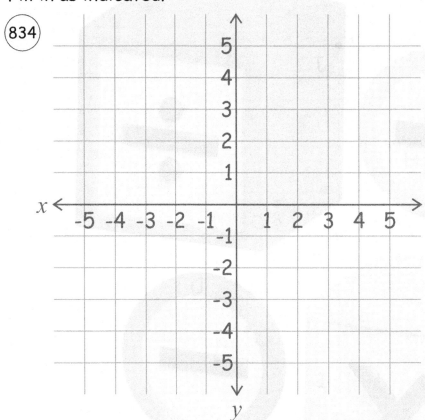

(834)

A = (5, -2)   B = (3, 2)

C = (-4, 4)   D = (3, 1)

E = (4, 1)   F = (-2, 3)

G = (-4, 2)   H = (-5, -3)

I = (-5, -4)   J = (-3, 4)

Cartesian Coordinates With Four Quadrants

Fill in as indicated.

(835)

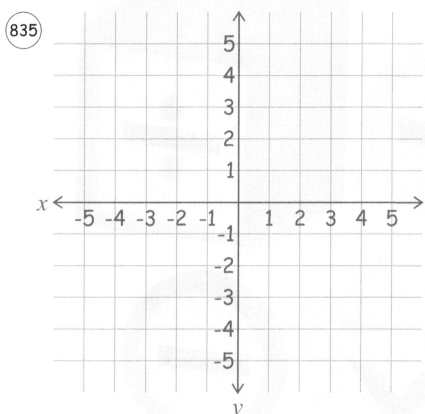

A = (-4, -3) B = (1, 5)

C = (3, 2) D = (-5, -4)

E = (-4, 2) F = (4, 1)

G = (-3, -2) H = (3, 3)

I = (3, -3) J = (-1, -2)

Cartesian Coordinates With Four Quadrants

Fill in as indicated.

(836)

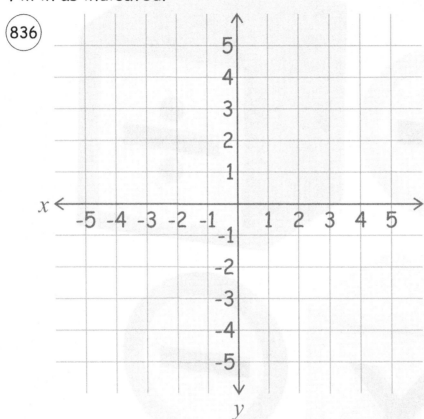

A = (5, -3)   B = (3, 2)

C = (-2, 1)   D = (2, -3)

E = (-5, -2)   F = (2, -5)

G = (-4, 1)   H = (-4, -3)

I = (-4, -5)   J = (1, 3)

Cartesian Coordinates With Four Quadrants

Fill in as indicated.

(837)

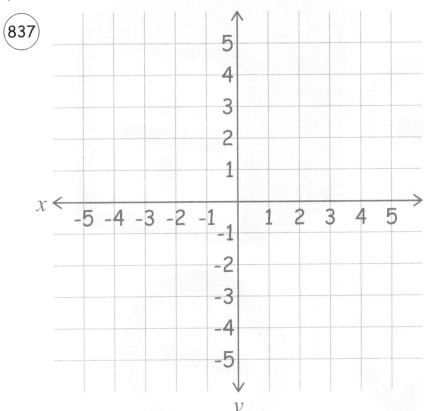

A = (4, -4)   B = (0, 2)

C = (5, 0)   D = (0, -3)

E = (-1, 5)   F = (0, 4)

G = (-3, -2)   H = (-4, -5)

I = (-1, 0)   J = (2, 4)

Cartesian Coordinates With Four Quadrants

Fill in as indicated.

(838)

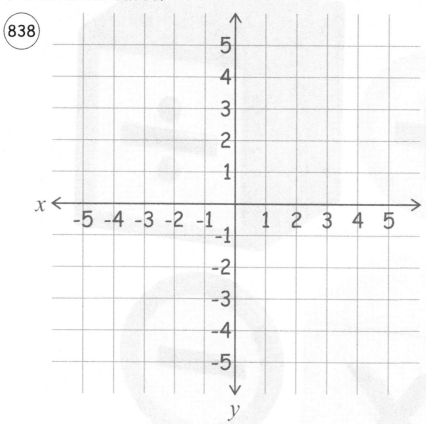

A = (5, 3)   B = (0, 2)

C = (-4, 1)   D = (2, 3)

E = (2, 1)   F = (0, 3)

G = (3, -4)   H = (3, 3)

I = (-5, 5)   J = (5, 2)

Plot Lines

Plot and draw the lines.

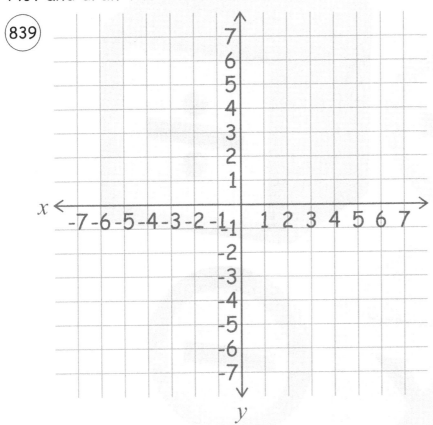

(839)

A = (-1, 6)   B = (4, 1)

C = (7, -2)   D = (-2, 7)

E = (3, 2)   F = (5, 0)

Plot Lines

Plot and draw the lines.

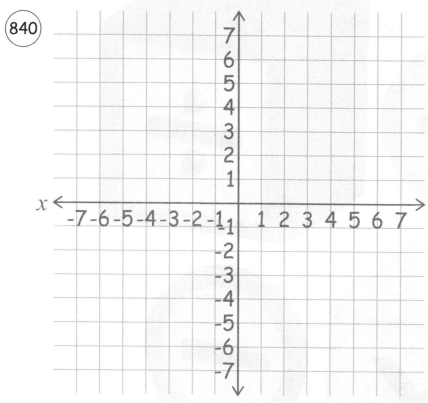

(840)

A = (-1, -7)   B = (4, -2)

C = (3, -3)   D = (2, -4)

E = (0, -6)   F = (1, -5)

Plot Lines

Plot and draw the lines.

(841)

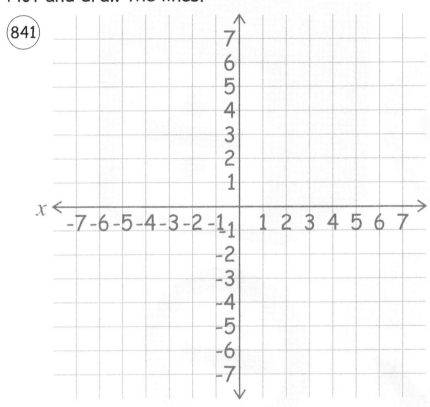

A = (-2, 0)   B = (4, -6)

C = (1, -3)   D = (5, -7)

E = (-5, 3)   F = (-7, 5)

Name: _ _ _ _ _ _ _ _ _ _
Date: __/__/____
Time Taken: ____ Min

Plot Lines

Plot and draw the lines.

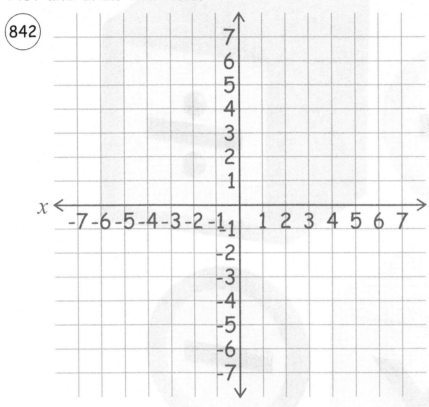

(842)

A = (-6, 2)   B = (1, 2)

C = (-7, 2)   D = (-1, 2)

E = (4, 2)   F = (2, 2)

Plot Lines

Plot and draw the lines.

(843)

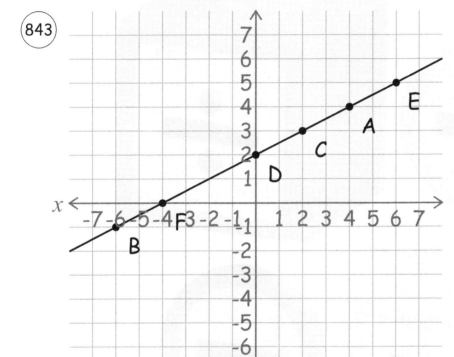

A = _____   B = _____

C = _____   D = _____

E = _____   F = _____

Plot Lines

Plot and draw the lines.

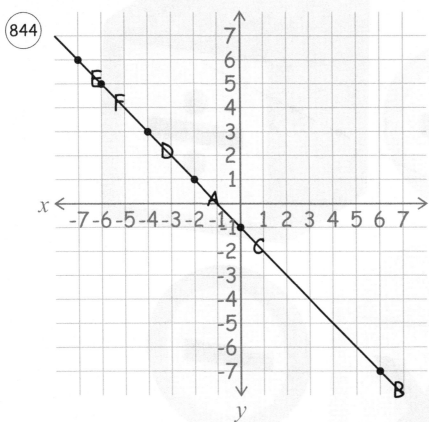

A = _____   B = _____

C = _____   D = _____

E = _____   F = _____

Plot Lines

Plot and draw the lines.

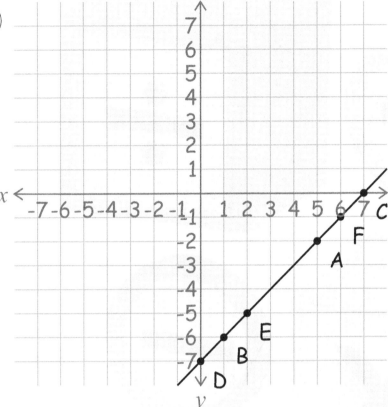

A = _____   B = _____

C = _____   D = _____

E = _____   F = _____

Plot Lines

Plot and draw the lines.

(846)

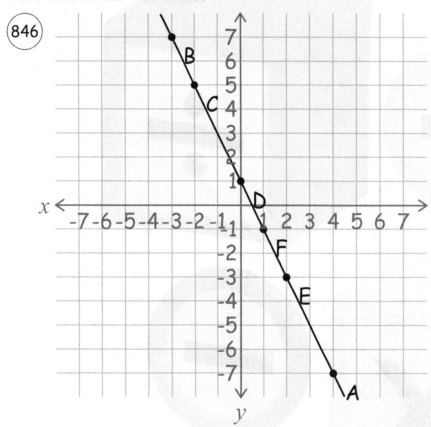

A = _____   B = _____

C = _____   D = _____

E = _____   F = _____

Name: _____

Date: __/__/____

Time Taken: ____ Min

Exponents

Convert the values.

(847) $53^2 =$ _____

(848) $20^3 =$ _____

(849) $40^2 =$ _____

(850) $72^3 =$ _____

(851) $50^3 =$ _____

(852) $53^3 =$ _____

(853) $89^3 =$ _____

(854) $95^3 =$ _____

(855) $89^2 =$ _____

(856) $88^2 =$ _____

(857) $67^3 =$ _____

(858) $75^2 =$ _____

(859) $52^3 =$ _____

(860) $93^2 =$ _____

(861) $12^3 =$ _____

(862) $23^3 =$ _____

(863) $99^3 =$ _____

(864) $85^2 =$ _____

(865) $66^3 =$ _____

(866) $55^3 =$ _____

Exponents

Convert the values.

(867) $38^3 =$ _____

(868) $17^3 =$ _____

(869) $43^2 =$ _____

(870) $25^2 =$ _____

(871) $79^2 =$ _____

(872) $55^2 =$ _____

(873) $7^3 =$ _____

(874) $87^3 =$ _____

(875) $98^2 =$ _____

(876) $95^2 =$ _____

(877) $52^2 =$ _____

(878) $92^3 =$ _____

(879) $3^2 =$ _____

(880) $21^3 =$ _____

(881) $44^3 =$ _____

(882) $4^2 =$ _____

(883) $20^3 =$ _____

(884) $56^3 =$ _____

(885) $95^3 =$ _____

(886) $41^3 =$ _____

Exponents

Convert the values.

(887) $62^2 =$ _____

(888) $57^3 =$ _____

(889) $69^3 =$ _____

(890) $6^3 =$ _____

(891) $83^2 =$ _____

(892) $49^3 =$ _____

(893) $84^3 =$ _____

(894) $81^3 =$ _____

(895) $70^2 =$ _____

(896) $30^3 =$ _____

(897) $56^3 =$ _____

(898) $1^2 =$ _____

(899) $81^2 =$ _____

(900) $51^2 =$ _____

(901) $17^3 =$ _____

(902) $91^2 =$ _____

(903) $52^3 =$ _____

(904) $12^3 =$ _____

(905) $72^3 =$ _____

(906) $77^3 =$ _____

Exponents

Convert the values.

(907) $60^2 =$ _____

(908) $99^2 =$ _____

(909) $41^3 =$ _____

(910) $74^3 =$ _____

(911) $52^3 =$ _____

(912) $29^2 =$ _____

(913) $45^2 =$ _____

(914) $70^2 =$ _____

(915) $95^2 =$ _____

(916) $86^2 =$ _____

(917) $98^3 =$ _____

(918) $68^3 =$ _____

(919) $93^3 =$ _____

(920) $30^3 =$ _____

(921) $55^2 =$ _____

(922) $92^3 =$ _____

(923) $55^3 =$ _____

(924) $32^3 =$ _____

(925) $98^2 =$ _____

(926) $99^3 =$ _____

Name:

Date:
__/__/____

Time Taken:
_____ Min

Scientific Notation

Provide the scientific notation for each value.

(927) 2,350,000 = _____   (928) 2,500,000 = _____

(929) 256,000 = _____   (930) 8,700,000 = _____

(931) 9,210,000 = _____   (932) 4,100,000 = _____

(933) 4,390,000 = _____   (934) 8,510,000 = _____

(935) 9,500,000 = _____   (936) 3,500,000 = _____

(937) 2,600,000 = _____   (938) 9,800,000 = _____

(939) 1,084,000 = _____   (940) 4,180,000 = _____

(941) 8,735,000 = _____   (942) 5,620,000 = _____

(943) 3,530,000 = _____   (944) 7,625,000 = _____

(945) 4,936,000 = _____   (946) 2,577,000 = _____

Scientific Notation

Provide the scientific notation for each value.

(947) 8,320,000 = _____ (948) 9,400,000 = _____

(949) 3,250,000 = _____ (950) 4,956,000 = _____

(951) 6,940,000 = _____ (952) 6,000,000 = _____

(953) 6,286,000 = _____ (954) 7,355,000 = _____

(955) 6,600,000 = _____ (956) 5,400,000 = _____

(957) 1,650,000 = _____ (958) 6,800,000 = _____

(959) 7,891,000 = _____ (960) 8,290,000 = _____

(961) 7,000,000 = _____ (962) 950,000 = _____

(963) 8,340,000 = _____ (964) 1,883,000 = _____

(965) 3,330,000 = _____ (966) 7,810,000 = _____

Scientific Notation

Provide the scientific notation for each value.

(967) 4,350,000 = _____   (968) 7,160,000 = _____

(969) 1,890,000 = _____   (970) 9,510,000 = _____

(971) 8,708,000 = _____   (972) 870,000 = _____

(973) 3,200,000 = _____   (974) 7,030,000 = _____

(975) 6,993,000 = _____   (976) 8,440,000 = _____

(977) 6,710,000 = _____   (978) 1,270,000 = _____

(979) 2,370,000 = _____   (980) 3,800,000 = _____

(981) 5,360,000 = _____   (982) 6,200,000 = _____

(983) 3,622,000 = _____   (984) 6,590,000 = _____

(985) 7,400,000 = _____   (986) 9,400,000 = _____

Scientific Notation

Provide the scientific notation for each value.

(987) $8.8 \times 10^6 =$ _____

(988) $9.624 \times 10^6 =$ _____

(989) $8.6 \times 10^4 =$ _____

(990) $7.57 \times 10^6 =$ _____

(991) $7.825 \times 10^6 =$ _____

(992) $1.75 \times 10^6 =$ _____

(993) $5 \times 10^6 =$ _____

(994) $9.357 \times 10^6 =$ _____

(995) $4.6 \times 10^6 =$ _____

(996) $3.025 \times 10^6 =$ _____

(997) $9.536 \times 10^6 =$ _____

(998) $9.2 \times 10^5 =$ _____

(999) $1.727 \times 10^6 =$ _____

(1000) $9.87 \times 10^6 =$ _____

(1001) $6.3 \times 10^6 =$ _____

(1002) $8.7 \times 10^6 =$ _____

(1003) $6.04 \times 10^6 =$ _____

(1004) $6.9 \times 10^6 =$ _____

(1005) $5.9 \times 10^6 =$ _____

(1006) $7.652 \times 10^6 =$ _____

Scientific Notation

Provide the scientific notation for each value.

(1007) $3.603 \times 10^6 =$ _____

(1008) $6.8 \times 10^6 =$ _____

(1009) $8.1 \times 10^6 =$ _____

(1010) $2.2 \times 10^4 =$ _____

(1011) $1.13 \times 10^6 =$ _____

(1012) $8.4 \times 10^6 =$ _____

(1013) $4.9 \times 10^6 =$ _____

(1014) $1.075 \times 10^6 =$ _____

(1015) $7.2 \times 10^6 =$ _____

(1016) $7.57 \times 10^6 =$ _____

(1017) $1.68 \times 10^6 =$ _____

(1018) $1.84 \times 10^6 =$ _____

(1019) $6.5 \times 10^6 =$ _____

(1020) $6.87 \times 10^6 =$ _____

(1021) $7.3 \times 10^5 =$ _____

(1022) $5.262 \times 10^6 =$ _____

(1023) $9 \times 10^6 =$ _____

(1024) $4.47 \times 10^6 =$ _____

(1025) $3.2 \times 10^6 =$ _____

(1026) $9.481 \times 10^6 =$ _____

Scientific Notation

Provide the scientific notation for each value.

(1027) $3.83 \times 10^6 =$ _____

(1028) $1.4 \times 10^5 =$ _____

(1029) $7.03 \times 10^6 =$ _____

(1030) $9 \times 10^6 =$ _____

(1031) $3.639 \times 10^6 =$ _____

(1032) $4.97 \times 10^6 =$ _____

(1033) $7.33 \times 10^6 =$ _____

(1034) $5.464 \times 10^6 =$ _____

(1035) $5.3 \times 10^6 =$ _____

(1036) $4.543 \times 10^6 =$ _____

(1037) $8.134 \times 10^6 =$ _____

(1038) $9.9 \times 10^5 =$ _____

(1039) $3.7 \times 10^6 =$ _____

(1040) $6.289 \times 10^6 =$ _____

(1041) $6.26 \times 10^6 =$ _____

(1042) $2.944 \times 10^6 =$ _____

(1043) $9.87 \times 10^6 =$ _____

(1044) $6.4 \times 10^6 =$ _____

(1045) $9.2 \times 10^6 =$ _____

(1046) $2 \times 10^6 =$ _____

Expressions - Single Step

Solve for the variable.

1047. $x + 1 = 8$ _____

1048. $x + 5 = 9$ _____

1049. $x + 2 = 9$ _____

1050. $1 - x = 0$ _____

1051. $x + 9 = 13$ _____

1052. $x + 2 = 10$ _____

1053. $9 + x = 11$ _____

1054. $5 + x = 9$ _____

1055. $x - 7 = -3$ _____

1056. $9 + x = 18$ _____

1057. $x - 1 = 5$ _____

1058. $x + 4 = 11$ _____

1059. $x + 9 = 15$ _____

1060. $x - 9 = -4$ _____

1061. $x + 1 = 4$ _____

1062. $x + 5 = 12$ _____

1063. $x - 9 = -8$ _____

1064. $2 - x = -4$ _____

1065. $x - 5 = -3$ _____

1066. $8 + x = 11$ _____

Expressions - Single Step

Solve for the variable.

(1067) $5 + x = 6$ _____

(1068) $x + 5 = 8$ _____

(1069) $6 - x = 2$ _____

(1070) $x - 7 = 0$ _____

(1071) $2 + x = 3$ _____

(1072) $3 - x = -5$ _____

(1073) $3 + x = 4$ _____

(1074) $x - 2 = 5$ _____

(1075) $1 - x = -4$ _____

(1076) $x - 9 = -8$ _____

(1077) $1 - x = 0$ _____

(1078) $x + 3 = 10$ _____

(1079) $x + 3 = 11$ _____

(1080) $x - 4 = -2$ _____

(1081) $x + 4 = 7$ _____

(1082) $x - 2 = 0$ _____

(1083) $8 - x = 4$ _____

(1084) $x + 6 = 10$ _____

(1085) $x - 4 = 3$ _____

(1086) $8 - x = 5$ _____

Expressions - Single Step

Solve for the variable.

(1087) $x - 9 = -5$ _____

(1088) $x - 4 = 2$ _____

(1089) $x - 7 = -6$ _____

(1090) $2 - x = -1$ _____

(1091) $6 - x = 5$ _____

(1092) $x + 7 = 9$ _____

(1093) $x + 6 = 13$ _____

(1094) $9 - x = 3$ _____

(1095) $5 + x = 9$ _____

(1096) $x + 9 = 16$ _____

(1097) $x - 1 = 8$ _____

(1098) $6 + x = 12$ _____

(1099) $5 + x = 11$ _____

(1100) $7 - x = 4$ _____

(1101) $7 - x = 1$ _____

(1102) $x + 7 = 15$ _____

(1103) $8 - x = 4$ _____

(1104) $x + 1 = 9$ _____

(1105) $x + 3 = 11$ _____

(1106) $9 - x = 2$ _____

Expressions - Single Step

Solve for the variable.

(1107) $4 + x = 9$ _____

(1108) $1 - x = -8$ _____

(1109) $1 + x = 8$ _____

(1110) $7 - x = 2$ _____

(1111) $9 - x = 1$ _____

(1112) $x + 4 = 6$ _____

(1113) $3 + x = 12$ _____

(1114) $2 + x = 4$ _____

(1115) $9 - x = 7$ _____

(1116) $x - 9 = -1$ _____

(1117) $4 + x = 7$ _____

(1118) $x - 2 = 3$ _____

(1119) $x + 8 = 17$ _____

(1120) $x + 8 = 9$ _____

(1121) $5 - x = 1$ _____

(1122) $3 - x = 2$ _____

(1123) $8 + x = 12$ _____

(1124) $9 - x = 3$ _____

(1125) $7 + x = 9$ _____

(1126) $3 + x = 6$ _____

Expressions - Single Step

Solve for the variable.

(1127) $7 - x = -1$ _____

(1128) $4 + x = 12$ _____

(1129) $2 + x = 3$ _____

(1130) $x + 4 = 6$ _____

(1131) $6 + x = 10$ _____

(1132) $6 - x = 2$ _____

(1133) $x - 1 = 2$ _____

(1134) $4 - x = 2$ _____

(1135) $x - 1 = 6$ _____

(1136) $x - 6 = -2$ _____

(1137) $x - 3 = -2$ _____

(1138) $1 + x = 2$ _____

(1139) $6 + x = 11$ _____

(1140) $x + 3 = 4$ _____

(1141) $7 + x = 13$ _____

(1142) $2 + x = 4$ _____

(1143) $2 - x = -5$ _____

(1144) $9 + x = 13$ _____

(1145) $7 + x = 8$ _____

(1146) $8 - x = 5$ _____

Name:

Date:
__/__/____
Time Taken:
____ Min

Number Problems

Solve.

(1147) ___ One-half of a number is 12. Find the number.

(1148) ___ A number decreased by 17 is 8. Find the number.

(1149) ___ Thirty more than a number is 46. What is the number?

(1150) ___ The sum of a number and eight is 34. Find the number.

(1151) ___ Seven less than a number is 12. Find the number.

(1152) ___ Twenty-seven less than a number is 27. Find the number.

(1153) ___ The sum of a number and four is 34. Find the number.

(1154) ___ Ten more than a number is 21. What is the number?

(1155) ___ A number decreased by 11 is 27. Find the number.

(1156) ___ One-third of a number is 2. Find the number.

Name: _____

Date: __/__/____

Time Taken: ____ Min

Number Problems

Solve.

(1157) ___ The sum of a number and 26 is 41. Find the number.

(1158) ___ A number increased by 21 is 44. Find the number.

(1159) ___ One-third of a number is 7. Find the number.

(1160) ___ A number increased by 23 is 47. Find the number.

(1161) ___ One-fifth of a number is 5. Find the number.

(1162) ___ A number increased by 22 is 43. Find the number.

(1163) ___ Three-fifths of a number is 12. Find the number.

(1164) ___ A number diminished by 12 is 22. Find the number.

(1165) ___ A number decreased by 27 is 9. Find the number.

(1166) ___ A number decreased by 11 is 7. Find the number.

Number Problems

Solve.

(1167) ____ Twenty-seven less than a number is 6. Find the number.

(1168) ____ A number decreased by 4 is 8. Find the number.

(1169) ____ A number diminished by 3 is 26. Find the number.

(1170) ____ A number decreased by 17 is 15. Find the number.

(1171) ____ Fifteen less than a number is 2. Find the number.

(1172) ____ Twenty-two less than a number is 5. Find the number.

(1173) ____ The sum of a number and 23 is 30. Find the number.

(1174) ____ A number increased by 15 is 37. Find the number.

(1175) ____ Twelve less than a number is 19. Find the number.

(1176) ____ A number increased by 20 is 44. Find the number.

Pre-Algebra Equations (One Step) Addition and Subtraction

Solve for the variable.

1177 $7 - x = 4$ _____

1178 $x + 1 = 10$ _____

1179 $6x + 2 = 38$ _____

1180 $x + 2 = 9$ _____

1181 $x + 8 = 10$ _____

1182 $5x + 5 = 15$ _____

1183 $5 + 4x = 33$ _____

1184 $4 - x = 3$ _____

1185 $2 + 6x = 26$ _____

1186 $7x + 1 = 15$ _____

1187 $8x - 4 = 28$ _____

1188 $17 - 3x = 2$ _____

1189 $8 + x = 14$ _____

1190 $x - 3 = 4$ _____

1191 $3 + x = 9$ _____

1192 $6 + 2x = 22$ _____

1193 $9 - x = 6$ _____

1194 $1x - 4 = 1$ _____

1195 $x + 3 = 11$ _____

1196 $4 + x = 7$ _____

Pre-Algebra Equations (One Step) Addition and Subtraction

Solve for the variable.

(1197) $5 + x = 11$ _____   (1198) $3 + 7x = 24$ _____

(1199) $9x + 9 = 63$ _____   (1200) $6x + 8 = 56$ _____

(1201) $x + 8 = 10$ _____   (1202) $2x - 6 = 6$ _____

(1203) $x + 2 = 5$ _____   (1204) $x - 5 = 4$ _____

(1205) $14 - 6x = 2$ _____   (1206) $38 - 5x = 3$ _____

(1207) $7 - x = 2$ _____   (1208) $5x + 7 = 22$ _____

(1209) $x - 4 = 1$ _____   (1210) $2x - 3 = 7$ _____

(1211) $81 - 8x = 9$ _____   (1212) $x - 1 = 8$ _____

(1213) $8x - 6 = 10$ _____   (1214) $x - 1 = 2$ _____

(1215) $59 - 8x = 3$ _____   (1216) $x + 4 = 13$ _____

Pre-Algebra Equations (One Step) Addition and Subtraction

Solve for the variable.

(1217) $47 - 8x = 7$ _____

(1218) $x - 6 = 1$ _____

(1219) $9 - x = 3$ _____

(1220) $4x - 8 = 12$ _____

(1221) $4 + x = 13$ _____

(1222) $x + 4 = 8$ _____

(1223) $x - 5 = 3$ _____

(1224) $2 + x = 10$ _____

(1225) $1 + x = 10$ _____

(1226) $5x + 5 = 20$ _____

(1227) $x + 7 = 8$ _____

(1228) $x - 2 = 4$ _____

(1229) $6 - x = 1$ _____

(1230) $x - 3 = 4$ _____

(1231) $5 + x = 6$ _____

(1232) $x - 3 = 6$ _____

(1233) $6x - 3 = 21$ _____

(1234) $8x + 2 = 74$ _____

(1235) $4x - 8 = 4$ _____

(1236) $8 - x = 2$ _____

Pre-Algebra Equations (One Step) Addition and Subtraction

Solve for the variable.

(1237) $4 - x = 2$ _____

(1238) $x + 3 = 12$ _____

(1239) $9x - 9 = 18$ _____

(1240) $6x + 7 = 25$ _____

(1241) $48 - 8x = 8$ _____

(1242) $3x + 1 = 4$ _____

(1243) $4x - 6 = 18$ _____

(1244) $69 - 8x = 5$ _____

(1245) $4x + 2 = 22$ _____

(1246) $36 - 9x = 0$ _____

(1247) $4 + x = 7$ _____

(1248) $x - 4 = 4$ _____

(1249) $1 + 7x = 36$ _____

(1250) $9 + 5x = 34$ _____

(1251) $35 - 7x = 7$ _____

(1252) $4x - 2 = 6$ _____

(1253) $1 + 8x = 17$ _____

(1254) $4x + 2 = 26$ _____

(1255) $x + 9 = 15$ _____

(1256) $2x - 9 = 5$ _____

Name:

Date: __/__/____

Time Taken: ____ Min

Pre-Algebra Equations (One Step) Addition and Subtraction

Solve for the variable.

(1257) x – 6 = 3 _____

(1258) 66 – 7x = 3 _____

(1259) 7 – x = 6 _____

(1260) 9 + 9x = 54 _____

(1261) x – 2 = 0 _____

(1262) 23 – 7x = 2 _____

(1263) x – 3 = 3 _____

(1264) x + 3 = 7 _____

(1265) x + 6 = 10 _____

(1266) x – 4 = 4 _____

(1267) 1x – 8 = 1 _____

(1268) 2 + x = 7 _____

(1269) x + 5 = 13 _____

(1270) 28 – 3x = 4 _____

(1271) 12 – 2x = 4 _____

(1272) 33 – 3x = 9 _____

(1273) 8x – 6 = 50 _____

(1274) 3x – 6 = 9 _____

(1275) 7 + 4x = 31 _____

(1276) 8x – 1 = 39 _____

Pre-Algebra Equations (One Step) Addition and Subtraction

Solve for the variable.

(1277) $8 + 1x = 17$ _____

(1278) $62 - 8x = 6$ _____

(1279) $7x - 5 = 30$ _____

(1280) $x + 8 = 15$ _____

(1281) $x - 6 = 1$ _____

(1282) $8 + x = 13$ _____

(1283) $4x - 1 = 19$ _____

(1284) $2 + x = 7$ _____

(1285) $x - 4 = 4$ _____

(1286) $x + 5 = 8$ _____

(1287) $x - 4 = 0$ _____

(1288) $5 + 1x = 12$ _____

(1289) $6 - x = 4$ _____

(1290) $43 - 5x = 8$ _____

(1291) $6 - x = 1$ _____

(1292) $5 + 4x = 21$ _____

(1293) $14 - 6x = 2$ _____

(1294) $6 + 5x = 51$ _____

(1295) $8 - x = 7$ _____

(1296) $5 - x = 2$ _____

Pre-Algebra Equations (One Step) Addition and Subtraction

Solve for the variable.

(1297) $1 + x = 2$ _____

(1298) $3x - 2 = 25$ _____

(1299) $5x - 2 = 18$ _____

(1300) $x - 3 = 6$ _____

(1301) $5 - 1x = 0$ _____

(1302) $14 - 2x = 2$ _____

(1303) $5x + 5 = 10$ _____

(1304) $2 + x = 8$ _____

(1305) $7x + 3 = 59$ _____

(1306) $x - 7 = 2$ _____

(1307) $x - 5 = 2$ _____

(1308) $3x - 3 = 3$ _____

(1309) $6x - 8 = 28$ _____

(1310) $8 - 1x = 1$ _____

(1311) $5 + x = 7$ _____

(1312) $17 - 6x = 5$ _____

(1313) $9 - x = 4$ _____

(1314) $7 + 6x = 25$ _____

(1315) $x + 1 = 10$ _____

(1316) $3 - x = 2$ _____

Pre-Algebra Equations (One Step) Addition and Subtraction

Solve for the variable.

(1317) $8x + 4 = 28$ _____

(1318) $12 - 1x = 3$ _____

(1319) $24 - 3x = 6$ _____

(1320) $27 - 3x = 0$ _____

(1321) $9 + x = 14$ _____

(1322) $9x - 5 = 67$ _____

(1323) $9 + 2x = 27$ _____

(1324) $5x - 6 = 19$ _____

(1325) $x - 2 = 1$ _____

(1326) $54 - 9x = 9$ _____

(1327) $x - 1 = 5$ _____

(1328) $2 - x = 0$ _____

(1329) $x + 4 = 6$ _____

(1330) $6x - 4 = 20$ _____

(1331) $8 + x = 13$ _____

(1332) $4 - x = 3$ _____

(1333) $7x - 6 = 22$ _____

(1334) $x + 8 = 17$ _____

(1335) $4x + 1 = 13$ _____

(1336) $2 + 5x = 22$ _____

Pre-Algebra Equations (One Step) Addition and Subtraction

Solve for the variable.

(1337) $x + 2 = 10$ _____   (1338) $2 + 1x = 5$ _____

(1339) $51 - 7x = 9$ _____   (1340) $3x - 9 = 3$ _____

(1341) $x - 8 = 1$ _____   (1342) $5x - 4 = 6$ _____

(1343) $6 - x = 4$ _____   (1344) $9x - 9 = 45$ _____

(1345) $9x + 8 = 44$ _____   (1346) $1 + 8x = 17$ _____

(1347) $8x + 7 = 39$ _____   (1348) $3 - x = 1$ _____

(1349) $x - 1 = 4$ _____   (1350) $7x - 7 = 0$ _____

(1351) $x - 4 = 5$ _____   (1352) $8x - 6 = 18$ _____

(1353) $75 - 8x = 3$ _____   (1354) $x - 3 = 3$ _____

(1355) $x + 9 = 14$ _____   (1356) $9 - x = 2$ _____

Pre-Algebra Equations (One Step) Addition and Subtraction

Solve for the variable.

(1357) $6x - 5 = 7$ _____

(1358) $8 + 8x = 24$ _____

(1359) $5 + 8x = 13$ _____

(1360) $8x - 9 = 47$ _____

(1361) $11 - 1x = 6$ _____

(1362) $4 - x = 0$ _____

(1363) $43 - 7x = 8$ _____

(1364) $6x - 7 = 29$ _____

(1365) $x - 3 = 1$ _____

(1366) $x - 1 = 6$ _____

(1367) $x + 4 = 11$ _____

(1368) $8 - x = 6$ _____

(1369) $3x - 7 = 14$ _____

(1370) $x + 4 = 12$ _____

(1371) $5x + 4 = 19$ _____

(1372) $8 - x = 3$ _____

(1373) $11 - 9x = 2$ _____

(1374) $x + 9 = 13$ _____

(1375) $9 + 3x = 24$ _____

(1376) $3 + 4x = 31$ _____

Pre-Algebra Equations (One Step) Multiplication and Division

Solve for the variable.

(1377) $x \div 9 = 1$ _____

(1378) $3 \times x = 27$ _____

(1379) $9x - 5 = 58$ _____

(1380) $4x + 8 = 32$ _____

(1381) $6 + 5x = 46$ _____

(1382) $x \times 6 = 6$ _____

(1383) $8x + 9 = 33$ _____

(1384) $x \times 8 = 64$ _____

(1385) $8 + 9x = 35$ _____

(1386) $x \times 1 = 9$ _____

(1387) $x \div 5 = 6$ _____

(1388) $21 \div x = 3$ _____

(1389) $12 - 9x = 3$ _____

(1390) $1 + 5x = 21$ _____

(1391) $8 - 4x = 4$ _____

(1392) $71 - 9x = 8$ _____

(1393) $7x + 6 = 27$ _____

(1394) $5x + 6 = 51$ _____

(1395) $9x + 6 = 42$ _____

(1396) $14 \div x = 7$ _____

Pre-Algebra Equations (One Step) Multiplication and Division

Solve for the variable.

1397 $x \div 9 = 7$ _____

1398 $5 \times x = 15$ _____

1399 $5 + 2x = 15$ _____

1400 $1x + 5 = 7$ _____

1401 $18 \div x = 6$ _____

1402 $24 \div x = 6$ _____

1403 $4 + 1x = 8$ _____

1404 $x \times 4 = 12$ _____

1405 $28 \div x = 7$ _____

1406 $56 \div x = 7$ _____

1407 $9x + 5 = 14$ _____

1408 $x \times 8 = 64$ _____

1409 $2 \times x = 18$ _____

1410 $8 \times x = 24$ _____

1411 $9x + 9 = 63$ _____

1412 $x \div 2 = 6$ _____

1413 $2 \div x = 1$ _____

1414 $x \div 9 = 9$ _____

1415 $4 + 7x = 39$ _____

1416 $8 + 2x = 20$ _____

Pre-Algebra Equations (One Step) Multiplication and Division

Solve for the variable.

1417 x × 6 = 24 _____

1418 6 + 2x = 24 _____

1419 x × 9 = 36 _____

1420 4 + 1x = 13 _____

1421 x ÷ 9 = 6 _____

1422 7 × x = 28 _____

1423 67 – 8x = 3 _____

1424 7x + 3 = 66 _____

1425 1 × x = 9 _____

1426 6 + 8x = 54 _____

1427 14 – 1x = 6 _____

1428 5x + 8 = 48 _____

1429 16 – 8x = 0 _____

1430 6 ÷ x = 3 _____

1431 30 ÷ x = 6 _____

1432 12 – 1x = 7 _____

1433 x ÷ 6 = 8 _____

1434 x × 2 = 8 _____

1435 7 × x = 14 _____

1436 3 – 1x = 2 _____

Pre-Algebra Equations (One Step) Multiplication and Division

Solve for the variable.

(1437) $x \div 6 = 5$ _____

(1438) $x \div 2 = 6$ _____

(1439) $x \times 4 = 36$ _____

(1440) $x \div 7 = 5$ _____

(1441) $3 \times x = 12$ _____

(1442) $3 + 3x = 6$ _____

(1443) $32 \div x = 4$ _____

(1444) $7x + 2 = 9$ _____

(1445) $47 - 7x = 5$ _____

(1446) $5x - 6 = 34$ _____

(1447) $4x - 2 = 30$ _____

(1448) $21 - 9x = 3$ _____

(1449) $63 \div x = 9$ _____

(1450) $24 \div x = 4$ _____

(1451) $x \times 4 = 12$ _____

(1452) $7x + 7 = 21$ _____

(1453) $1x - 6 = 0$ _____

(1454) $7 \times x = 49$ _____

(1455) $x \times 9 = 36$ _____

(1456) $3 \times x = 21$ _____

Pre-Algebra Equations (One Step) Multiplication and Division

Solve for the variable.

(1457) $56 - 6x = 2$ _____

(1458) $6x + 3 = 27$ _____

(1459) $x \div 6 = 4$ _____

(1460) $x \div 8 = 7$ _____

(1461) $x \div 2 = 7$ _____

(1462) $4x + 1 = 5$ _____

(1463) $16 - 8x = 0$ _____

(1464) $7x - 7 = 14$ _____

(1465) $5x - 3 = 22$ _____

(1466) $8 \times x = 64$ _____

(1467) $1x + 3 = 4$ _____

(1468) $2x + 7 = 17$ _____

(1469) $71 - 9x = 8$ _____

(1470) $6 \times x = 36$ _____

(1471) $x \div 5 = 8$ _____

(1472) $x \times 6 = 30$ _____

(1473) $2 \times x = 2$ _____

(1474) $x \times 7 = 28$ _____

(1475) $7x + 7 = 56$ _____

(1476) $1x + 1 = 3$ _____

Pre-Algebra Equations (One Step) Multiplication and Division

Solve for the variable.

(1477) $2 + 2x = 18$ _____

(1478) $3 + 7x = 24$ _____

(1479) $9 - 1x = 3$ _____

(1480) $x \times 8 = 8$ _____

(1481) $37 - 8x = 5$ _____

(1482) $3 \div x = 3$ _____

(1483) $11 - 5x = 6$ _____

(1484) $3x + 8 = 35$ _____

(1485) $x \times 7 = 35$ _____

(1486) $x \times 6 = 12$ _____

(1487) $9 \div x = 9$ _____

(1488) $3x - 8 = 1$ _____

(1489) $x \div 2 = 4$ _____

(1490) $18 \div x = 3$ _____

(1491) $x \times 1 = 9$ _____

(1492) $9x + 2 = 56$ _____

(1493) $6 \times x = 30$ _____

(1494) $7x - 5 = 30$ _____

(1495) $28 - 5x = 8$ _____

(1496) $45 \div x = 5$ _____

Pre-Algebra Equations (One Step) Multiplication and Division

Solve for the variable.

(1497) $x \times 4 = 8$ _____

(1498) $2 - 1x = 0$ _____

(1499) $x \times 8 = 8$ _____

(1500) $x \div 3 = 8$ _____

(1501) $x \times 7 = 42$ _____

(1502) $35 \div x = 5$ _____

(1503) $7 \times x = 21$ _____

(1504) $x \div 3 = 4$ _____

(1505) $8x - 2 = 6$ _____

(1506) $12 - 1x = 3$ _____

(1507) $3 \times x = 18$ _____

(1508) $46 - 9x = 1$ _____

(1509) $8 \times x = 24$ _____

(1510) $3 + 2x = 21$ _____

(1511) $9x - 1 = 80$ _____

(1512) $x \times 3 = 6$ _____

(1513) $x \times 1 = 4$ _____

(1514) $5 + 8x = 13$ _____

(1515) $15 - 7x = 1$ _____

(1516) $x \div 8 = 2$ _____

Pre-Algebra Equations (One Step) Multiplication and Division

Solve for the variable.

(1517) $6 \times x = 54$ _____

(1518) $x \times 5 = 25$ _____

(1519) $88 - 9x = 7$ _____

(1520) $7 \times x = 42$ _____

(1521) $26 - 8x = 2$ _____

(1522) $3 + 9x = 75$ _____

(1523) $24 \div x = 6$ _____

(1524) $1 \times x = 4$ _____

(1525) $65 - 7x = 2$ _____

(1526) $8x + 5 = 21$ _____

(1527) $7x - 6 = 29$ _____

(1528) $32 \div x = 4$ _____

(1529) $x \div 7 = 3$ _____

(1530) $2x - 1 = 1$ _____

(1531) $8x + 9 = 25$ _____

(1532) $48 \div x = 6$ _____

(1533) $9 \times x = 81$ _____

(1534) $18 - 3x = 3$ _____

(1535) $7x + 2 = 58$ _____

(1536) $9 + 6x = 45$ _____

Pre-Algebra Equations (One Step) Multiplication and Division

Solve for the variable.

(1537) $x \times 1 = 9$ _____

(1538) $4 \div x = 1$ _____

(1539) $4 \div x = 4$ _____

(1540) $13 - 5x = 8$ _____

(1541) $11 - 1x = 5$ _____

(1542) $17 - 2x = 1$ _____

(1543) $1x + 5 = 7$ _____

(1544) $26 - 8x = 2$ _____

(1545) $3 \times x = 9$ _____

(1546) $34 - 4x = 6$ _____

(1547) $5 \times x = 30$ _____

(1548) $x \div 5 = 4$ _____

(1549) $4x + 8 = 16$ _____

(1550) $72 \div x = 8$ _____

(1551) $x \div 3 = 9$ _____

(1552) $34 - 6x = 4$ _____

(1553) $x \times 2 = 14$ _____

(1554) $x \div 6 = 5$ _____

(1555) $x \times 9 = 45$ _____

(1556) $2x + 7 = 19$ _____

Pre-Algebra Equations (One Step) Multiplication and Division

Solve for the variable.

(1557) $16 - 2x = 2$ _____

(1558) $7 - 3x = 4$ _____

(1559) $2x - 1 = 17$ _____

(1560) $4 - 3x = 1$ _____

(1561) $8 + 9x = 53$ _____

(1562) $3x - 9 = 12$ _____

(1563) $8x - 3 = 45$ _____

(1564) $3 \times x = 9$ _____

(1565) $3 \times x = 3$ _____

(1566) $18 \div x = 6$ _____

(1567) $9x - 5 = 40$ _____

(1568) $5 + 8x = 37$ _____

(1569) $25 - 7x = 4$ _____

(1570) $15 \div x = 3$ _____

(1571) $x \div 5 = 9$ _____

(1572) $27 \div x = 3$ _____

(1573) $4x + 4 = 40$ _____

(1574) $5x + 2 = 22$ _____

(1575) $9x - 7 = 20$ _____

(1576) $7x - 2 = 54$ _____

Pre-Algebra Equations (One Step) Multiplication and Division

Solve for the variable.

1577 $1 + 7x = 8$ _____

1578 $x \div 8 = 9$ _____

1579 $9x - 6 = 39$ _____

1580 $2 + 6x = 38$ _____

1581 $1 \times x = 3$ _____

1582 $x \times 1 = 2$ _____

1583 $8x + 1 = 73$ _____

1584 $8x - 5 = 3$ _____

1585 $1 \times x = 8$ _____

1586 $x \times 6 = 42$ _____

1587 $x \times 4 = 32$ _____

1588 $9x - 2 = 7$ _____

1589 $8x - 1 = 47$ _____

1590 $8 + 2x = 10$ _____

1591 $14 \div x = 7$ _____

1592 $x \div 5 = 6$ _____

1593 $x \times 2 = 14$ _____

1594 $x \div 9 = 3$ _____

1595 $5 + 5x = 20$ _____

1596 $x \div 5 = 5$ _____

Pre-Algebra Equations (One Step) Multiplication and Division

Solve for the variable.

(1597) $1x + 5 = 8$ _____

(1598) $3 \times x = 18$ _____

(1599) $1 \times x = 3$ _____

(1600) $8 + 2x = 10$ _____

(1601) $42 \div x = 7$ _____

(1602) $9 \times x = 81$ _____

(1603) $6x + 6 = 24$ _____

(1604) $16 - 1x = 8$ _____

(1605) $4 \div x = 2$ _____

(1606) $x \div 4 = 8$ _____

(1607) $x \times 7 = 7$ _____

(1608) $1x - 5 = 0$ _____

(1609) $9 \times x = 9$ _____

(1610) $2 \times x = 8$ _____

(1611) $5 + 7x = 68$ _____

(1612) $64 - 7x = 1$ _____

(1613) $39 - 5x = 9$ _____

(1614) $56 \div x = 8$ _____

(1615) $x \div 4 = 1$ _____

(1616) $x \div 9 = 3$ _____

Pre-Algebra Equations (One Step) Multiplication and Division

Solve for the variable.

(1617) $x \times 4 = 16$ _____

(1618) $4 + 2x = 10$ _____

(1619) $1 \times x = 6$ _____

(1620) $x \div 7 = 9$ _____

(1621) $x \times 3 = 27$ _____

(1622) $6 + 8x = 30$ _____

(1623) $7x + 2 = 9$ _____

(1624) $7 - 7x = 0$ _____

(1625) $45 - 5x = 0$ _____

(1626) $18 \div x = 2$ _____

(1627) $5 \times x = 35$ _____

(1628) $x \times 9 = 72$ _____

(1629) $x \div 5 = 6$ _____

(1630) $9 - 8x = 1$ _____

(1631) $46 - 8x = 6$ _____

(1632) $2 + 8x = 42$ _____

(1633) $5 \times x = 20$ _____

(1634) $2x + 9 = 19$ _____

(1635) $4x - 8 = 4$ _____

(1636) $8 \times x = 72$ _____

Pre-Algebra Equations (Two Sides)

Solve for the variable.

(1637) $10 - x + 6 = 4 + 2x + 3$

(1638) $33 - x = 4x + 3$

(1639) $9 + 5x + 5 = 30 + x + 4$

(1640) $22 + 4x = 8x + 6$

(1641) $22 - x = 7 + 4x$

(1642) $59 - x = 5x + 5$

(1643) $3x = 16 - x$

(1644) $30 - x = 3x + 6$

(1645) $2 + 8x + 7 = 90 - x$

(1646) $94 - 7x = 4 + 8x$

Pre-Algebra Equations (Two Sides)

Solve for the variable.

1647 $23 + x = 2 + 4x$

1648 $112 - 5x = 8 + 8x$

1649 $3 + 5x = 27 + x$

1650 $6x + 8 = 2 + 7x$

1651 $98 - 4x = 2 + 8x$

1652 $40 + x + 0 = 5 + 7x + 5$

1653 $24 + x = 5x + 4$

1654 $9 + 7x + 9 = 51 + x + 3$

1655 $18 + x = 7x + 6$

1656 $60 - x = 4 + 7x + 8$

Pre-Algebra Equations (Two Sides)

Solve for the variable.

1657 $6x + 4 = 46 - x$

1658 $3x + 32 = 9x + 8$

1659 $5x + 8 = 50 - x$

1660 $100 - 5x = 9 + 8x$

1661 $7x + 5 = 35 + x$

1662 $27 + x = 4x$

1663 $8x = 81 - x$

1664 $23 + x = 5 + 7x$

1665 $5 + 6x = 77 - 2x$

1666 $6 + 3x = 12 + x$

Name: _____

Date: __/__/__

Time Taken: _____ Min

Pre-Algebra Equations (Two Sides)

Solve for the variable.

(1667) 20 - x + 5 = 3 + 8x + 4

(1668) 6x = 14 - x

(1669) 5 + 7x + 4 = 33 + x + 0

(1670) 4 + 2x + 8 = 39 - x

(1671) 8x + 5 = 41 - 4x

(1672) 50 + x = 4 + 8x + 4

(1673) 4x + 26 = 2 + 8x

(1674) 24 + x + 2 = 3 + 7x + 5

(1675) 24 + 7x = 9x + 8

(1676) 6 + 7x = 96 - 3x

Pre-Algebra Equations (Two Sides)

Solve for the variable.

(1677) $4 + 5x + 8 = 18 + x + 2$

(1678) $56 - x = 7x$

(1679) $5x + 6 = 42 + x$

(1680) $60 + x + -6 = 4 + 7x + 2$

(1681) $67 + x + 0 = 9 + 7x + 4$

(1682) $30 + x + -6 = 6 + 3x + 2$

(1683) $3 + 4x + 8 = 24 + x + -1$

(1684) $13 + x = 3 + 2x + 4$

(1685) $4x = 15 + x$

(1686) $3x = 28 - x$

Pre-Algebra Equations (Two Sides)

Solve for the variable.

1687) $2 + 5x + 2 = 32 + x$

1688) $9 + 3x + 5 = 26 - x$

1689) $4 + 4x + 3 = 16 + x + 0$

1690) $9 + x = 2x$

1691) $6x + 6 = 26 + x$

1692) $32 + x = 5 + 6x + 2$

1693) $3x + 7 = 37 - 2x$

1694) $9x + 2 = 155 - 8x$

1695) $17 - x = 7 + 3x + 2$

1696) $68 - x = 6x + 5$

Pre-Algebra Equations (Two Sides)

Solve for the variable.

(1697) $5x = 30 - x$

(1698) $59 + x = 9 + 8x + 8$

(1699) $4x = 35 - x$

(1700) $27 + x = 4x$

(1701) $10 + x = 3x$

(1702) $5x = 42 - x$

(1703) $7x = 48 + x$

(1704) $2x = 8 + x$

(1705) $7 + 4x = 17 - x$

(1706) $3x + 3 = 19 + x$

Name:
_____
Date:
__/__/___
Time Taken:
____ Min

Pre-Algebra Equations (Two Sides)

Solve for the variable.

(1707) $6 + 8x + 6 = 57 - x$

(1708) $8x + 4 = 7x + 9$

(1709) $5x = 12 - x$

(1710) $42 - x = 5x$

(1711) $6 + 8x = 24 + 2x$

(1712) $43 + x = 2 + 5x + 9$

(1713) $9x + 8 = 8x + 15$

(1714) $12 - x = 3x$

(1715) $3 + 5x = 12 + 4x$

(1716) $65 - x = 6x + 2$

Pre-Algebra Equations (Two Sides)

Solve for the variable.

(1717) 8 + 7x = 40 - x

(1718) 2 + 5x = 38 - x

(1719) 37 + x = 5x + 9

(1720) 3x + 5 = 41 - x

(1721) 2 + 4x = 18 + 2x

(1722) 71 - x = 6x + 8

(1723) 6x = 25 + x

(1724) 72 + x + -6 = 8 + 8x + 2

(1725) 53 - x = 5x + 5

(1726) 15 - x = 7 + 3x

Pre-Algebra Equations (Two Sides)

Solve for the variable.

1727 $4x + 3 = 21 + x$

1728 $3 + 5x + 6 = 29 + x$

1729 $42 + x = 7x$

1730 $27 + x = 4x$

1731 $16 + x = 2 + 3x$

1732 $5 + 5x + 9 = 26 - x$

1733 $8 + 8x = 13 + 7x$

1734 $7x = 54 + x$

1735 $10 + 3x = 7 + 4x$

1736 $3x = 28 - x$

Simplifying Expressions

1737 $4x - 4 + 4x - 9 + 7x + 6$

1738 $x + 3x$

1739 $-x + 3x$

1740 $-x + 1 + 2x + 5 + 3x - 5$

1741 $8x - 1 - x + 4$

1742 $7x + x$

1743 $4x + 8 - 4 - 2x + 5x$

Simplifying Expressions

(1744) x + x

(1745) 8x - x

(1746) -5 + 8x + 7 - 8x

(1747) 6x - x

(1748) 2x - 7x + 6 + 7

(1749) 7 + 6(-2x + 8)

(1750) 2 + 7(-7x + 6)

Simplifying Expressions

(1751) $2x + 2 + 9x + 9 + 9x + 3$

(1752) $6x - 4 - 6x + 5$

(1753) $-5x + 2x$

(1754) $9x + 2x$

(1755) $-x - 8x$

(1756) $9 + 8(9x + 6)$

(1757) $-9 - 8x + 2 - 4x$

Date:
__/__/____

Time Taken:
_____ Min

Simplifying Expressions

(1758) $x - 5x + 9x + 7 + 6$

(1759) $6x + 9 + 5x + 1 + 5x + 4$

(1760) $-x - x$

(1761) $-4 - x + 4 - 4x$

(1762) $1 + 7x - 8x$

(1763) $-x + 7x$

(1764) $2 + 9x + 4 + 5x$

Simplifying Expressions

(1765) $3x - x$

(1766) $1 + 9x - 7x$

(1767) $-9x + 8x + 1 - 7x$

(1768) $2x - 3x + 9x - 5 + 7$

(1769) $9x - 9x + 6 + 8$

(1770) $-6x + x$

(1771) $-3 + x - x - 8 + 3x$

Simplifying Expressions

(1772) $-8x - 7 + 2x$

(1773) $-8 + 8x - x - 3 - x$

(1774) $-5x + 6x + 9 - 2x$

(1775) $7 + 9x + 9 + 7x$

(1776) $5x + 5 - 8x - 8 + 4x - 7$

(1777) $6x + 4 + 2x + 1 + 8x + 7$

(1778) $-2x + 3 + 9x + 3 + x - 5$

Simplifying Expressions

(1779) -5x - 9x

(1780) 3 + 4(x - 1)

(1781) -8 - 8x + 2x - 8 + 7x

(1782) 3 + x - 1 + 7x

(1783) -9x + 4 + 7x + 4 + 8x - 7

(1784) -x + 7x

(1785) -4x + 8 - 7x

Simplifying Expressions

(1786) $-1 + 6 - 4x + 6x - 6 + 7x$

(1787) $8x + x$

(1788) $-9x + 5 - x$

(1789) $6x - x$

(1790) $3x + 1 + x + 2 + 2x + 3$

(1791) $5x + 4 + x$

(1792) $-9x + 8x + 8 - 8x$

Simplifying Expressions

(1793) -x - 4 - 9 - 4x

(1794) -7 + 4x + 7 - x

(1795) x + 7 + 2x

(1796) 4 + 5(6x - 9)

(1797) -9 - 9x + 7 - 3x

(1798) 7 + 1(4x - 4)

(1799) 7 + 8x + 8 + 4x

Simplifying Expressions

(1800) $1 + 9x + 2 + x$

(1801) $4x + 7 + x$

(1802) $3x + 7 + x$

(1803) $-8x + x$

(1804) $8 - 1(3x - 8)$

(1805) $5 + 4 + x - 7x + 1 - 5x$

(1806) $5x + x$

Simplifying Expressions

(1807) $4 + 6 + 7x - 7x + 4 - 3x$

(1808) $4x - 3 - 2x + 4$

(1809) $5 + 9 + 6x - 7x + 6 - 5x$

(1810) $5 + 4x - 8x + 9 - x$

(1811) $7x - 9x$

(1812) $-6 + 4x + 6 - 2x$

(1813) $1 + 7x - 9 + 2x$

Name:

Date:

Time Taken:
_____ Min

Simplifying Expressions

1814 $7x + 1 + x$

1815 $-8 + 8x - 9x - 7 - 4x$

1816 $-8 - 5x + 5 - 3x$

1817 $-5x + 7 - 3x$

1818 $6x + 5 + x$

1819 $x - x$

1820 $7x + x$

Inequalities - Addition and Subtraction

Solve.

1821 $6 \geq x + 1$

1822 $x - 7 \leq 8$

1823 $x - 2 \geq 6$

1824 $5 + x < 1$

1825 $7 - x < 9$

1826 $x + 1 \geq 9$

Inequalities - Addition and Subtraction

Solve.

1827
$x + 7 > 1$

1828
$9 > x - 7$

1829
$7 - x \leq 9$

1830
$x + 8 \leq 1$

1831
$3 + x < 9$

1832
$x - 1 \geq 5$

Inequalities - Addition and Subtraction

Solve.

1833 $3 \leq 3 + x$

1834 $x - 5 \geq 8$

1835 $2 + x < 6$

1836 $x - 7 > 9$

1837 $x + 8 \geq 1$

1838 $9 \geq 8 - x$

Inequalities - Multiplication and Division

Solve.

1839
$$\frac{x}{6} > 5$$

1840
$$9x \leq 12$$

1841
$$15 \leq 18x$$

1842
$$\frac{x}{2} \geq 3$$

1843
$$1 \geq \frac{x}{8}$$

1844
$$5x \leq 6$$

Name:
_____
Date:
__/__/____
Time Taken:
____ Min

Inequalities - Multiplication and Division

Solve.

(1845) $18\,x \geq 6$

(1846) $\dfrac{x}{5} \geq 6$

(1847) $\dfrac{x}{1} < 1$

(1848) $12\,x \leq 4$

(1849) $\dfrac{x}{2} > 3$

(1850) $8\,x \leq 10$

Inequalities - Multiplication and Division

Solve.

(1851) $\dfrac{x}{7} > 5$

(1852) $4\,x < 6$

(1853) $3\,x > 2$

(1854) $\dfrac{x}{1} > 7$

(1855) $\dfrac{x}{3} < 4$

(1856) $10 < 8\,x$

Find the Area and Perimeter

 (1857)

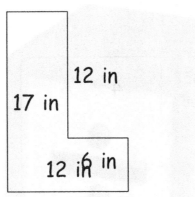

12 in
17 in
12 in 6 in

(1858)

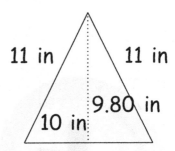

11 in   11 in
9.80 in
10 in

_____

(1859)

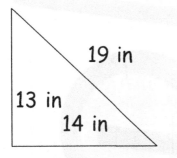

19 in
13 in
14 in

(1860)

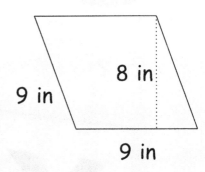

8 in
9 in
9 in

_____

(1861)

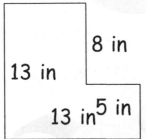

8 in
13 in
13 in 5 in

(1862)

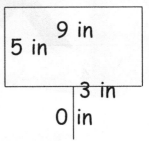

9 in
5 in
3 in
0 in

_____

Find the Area and Perimeter

(1863)

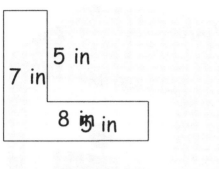

5 in
7 in
8 in 5 in

_____

(1864)

22 in

17 in

14 in

_____

(1865)

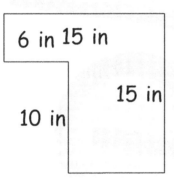

6 in 15 in

15 in

10 in

_____

(1866)

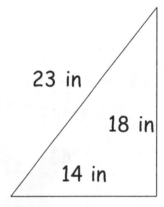

23 in

18 in

14 in

_____

(1867)

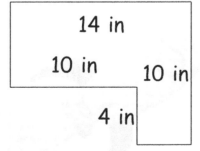

14 in

10 in

10 in

4 in

_____

(1868)

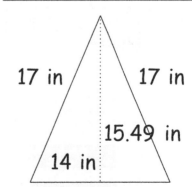

17 in   17 in

15.49 in

14 in

_____

Find the Area and Perimeter

(1869)

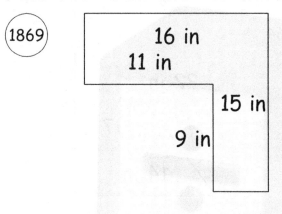

16 in
11 in
15 in
9 in

_____

(1870)

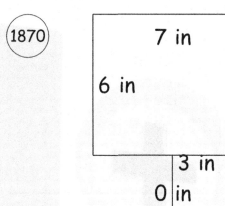

7 in
6 in
3 in
0 in

_____

(1871)

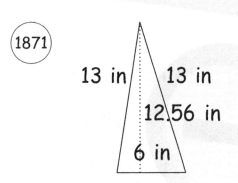

13 in   13 in
12.56 in
6 in

_____

(1872)

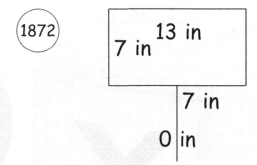

13 in
7 in
7 in
0 in

_____

(1873)

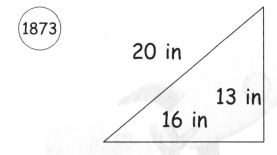

20 in
13 in
16 in

_____

(1874)

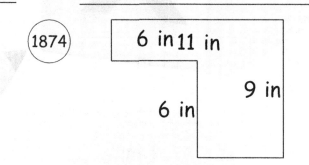

6 in 11 in
9 in
6 in

_____

Find the Volume and Surface Area

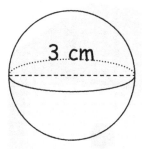

3 cm

_____

(1876)

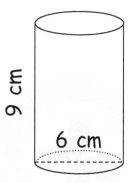

9 cm

6 cm

_____

(1877)

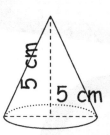

5 cm

5 cm

_____

(1878)

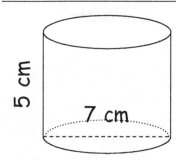

5 cm

7 cm

_____

(1879)

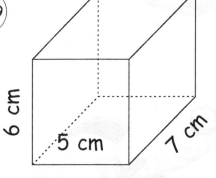

6 cm

5 cm

7 cm

_____

(1880)

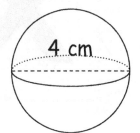

4 cm

_____

Find the Volume and Surface Area

(1881)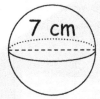

7 cm

_____

(1882)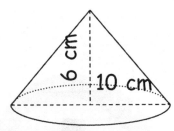

6 cm / 10 cm

_____

(1883)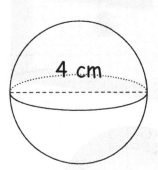

4 cm

_____

(1884)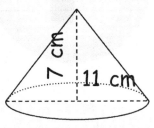

7 cm / 11 cm

_____

(1885)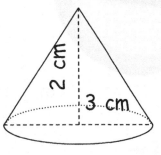

2 cm / 3 cm

_____

(1886)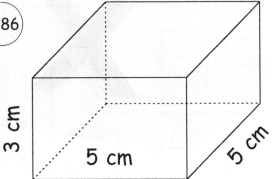

3 cm   5 cm   5 cm

_____

Fast Math Success Workbook Grade 6-7

148

Find the Volume and Surface Area

 1887

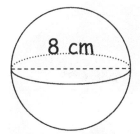

8 cm

_____

1888

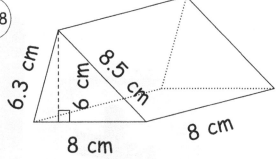

6.3 cm
6 cm
8.5 cm
8 cm
8 cm

_____

1889

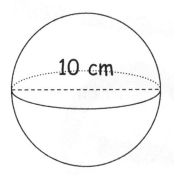

10 cm

_____

1890

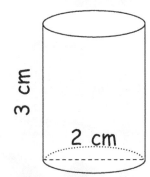

3 cm
2 cm

_____

1891

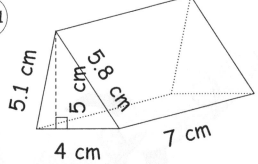

5.1 cm
5 cm
5.8 cm
4 cm
7 cm

_____

1892

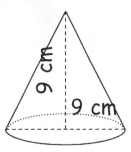

9 cm
9 cm

_____

Calculate the area of each circle.

(1893)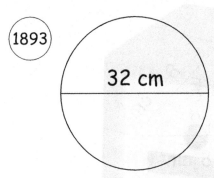
32 cm

_____

(1894)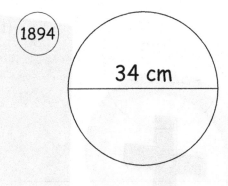
34 cm

_____

(1895)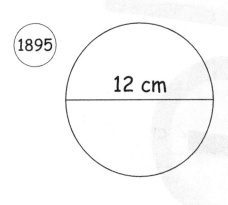
12 cm

_____

(1896)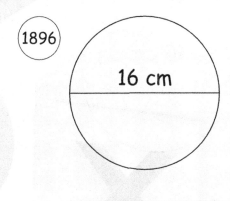
16 cm

_____

(1897)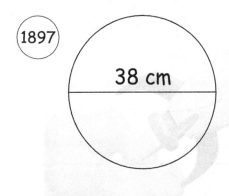
38 cm

_____

(1898)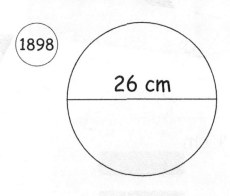
26 cm

_____

Calculate the area of each circle.

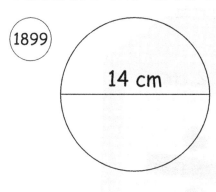

(1899) 14 cm

_____

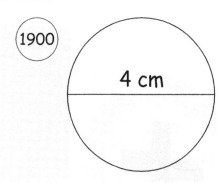

(1900) 4 cm

_____

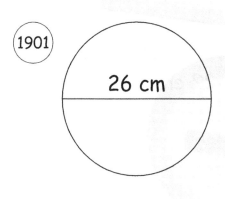

(1901) 26 cm

_____

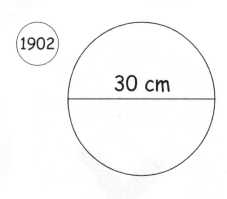

(1902) 30 cm

_____

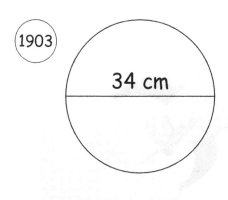

(1903) 34 cm

_____

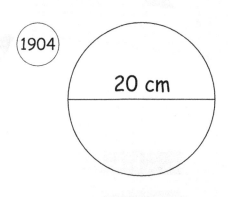

(1904) 20 cm

_____

Calculate the area of each circle.

(1905)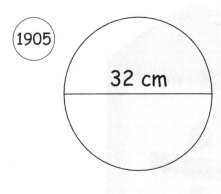

32 cm

_____

(1906)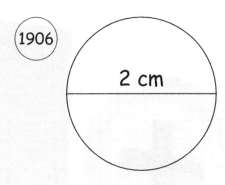

2 cm

_____

(1907)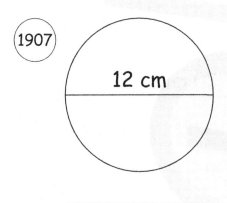

12 cm

_____

(1908)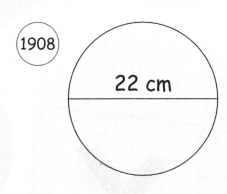

22 cm

_____

(1909)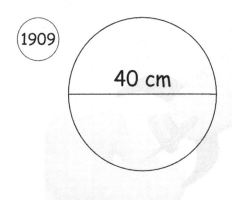

40 cm

_____

(1910)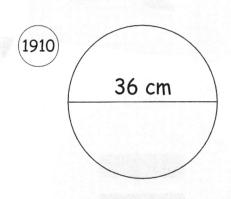

36 cm

_____

Calculate the circumference of each circle.

(1911)

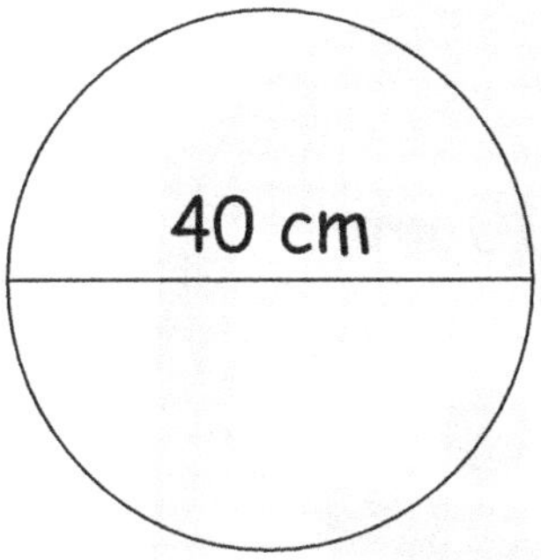

40 cm

_____

(1912)

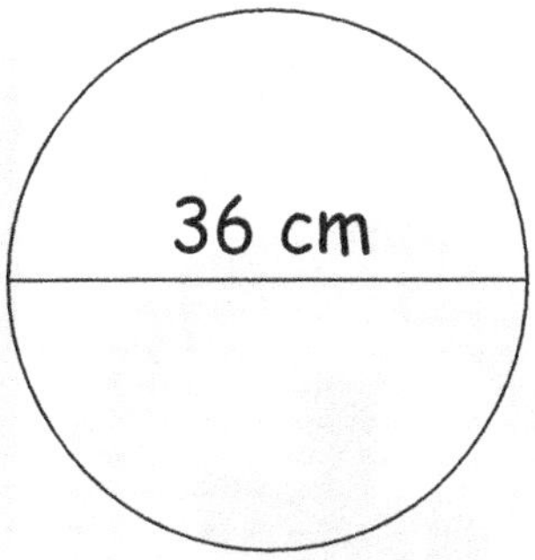

36 cm

_____

(1913)

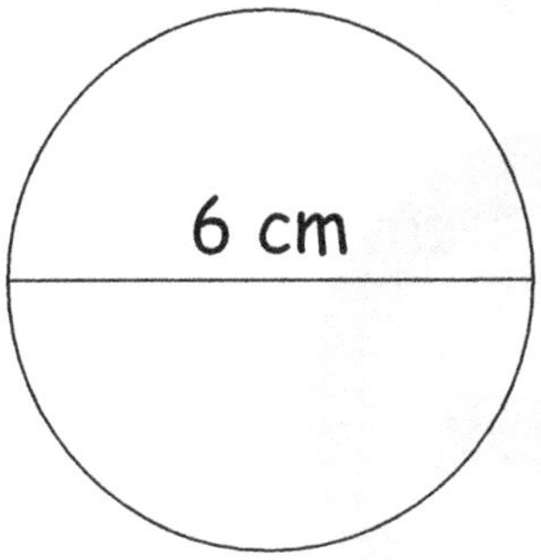

6 cm

_____

(1914)

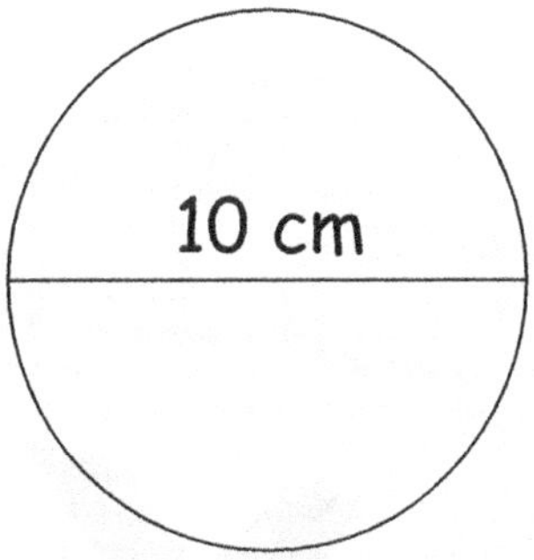

10 cm

_____

(1915)

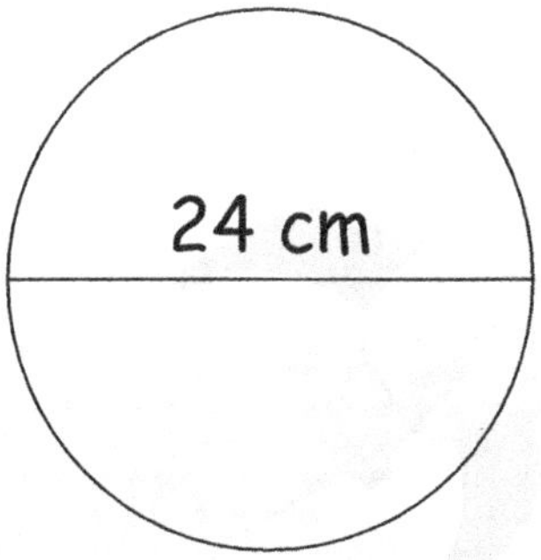

24 cm

_____

(1916)

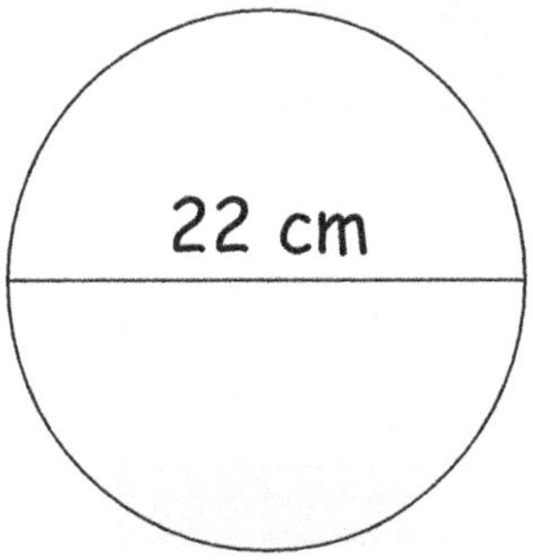

22 cm

_____

Calculate the circumference of each circle.

1917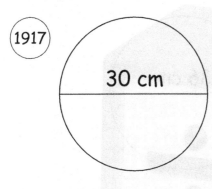

30 cm

_____

1918

10 cm

_____

1919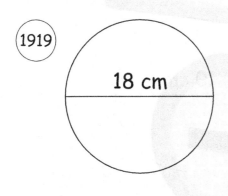

18 cm

_____

1920

4 cm

_____

1921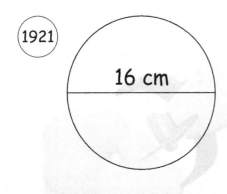

16 cm

_____

1922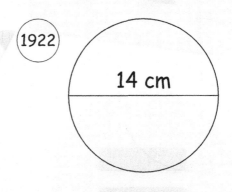

14 cm

_____

Calculate the circumference of each circle.

1923
36 cm
_____

1924
8 cm
_____

1925
38 cm
_____

1926
20 cm
_____

1927
22 cm
_____

1928
16 cm
_____

Measure of Center - Mean

Find the Mean of the following sets of data.

1929 29, 31, 55, 28, 71, 48, 38, 97, 89

Mean = ____

1930 67, 95, 50, 7, 6, 72

Mean = ____

1931 55, 67, 82, 63, 78, 67, 57

Mean = ____

1932 47, 74, 21, 50, 12, 80

Mean = _____

1933 41, 65, 62, 53, 35, 25

Mean = _____

1934 21, 89, 39, 85, 45, 86

Mean = _____

Name:
_____

Date:
__/__/____

Time Taken:
____ Min

Measure of Center - Mean

Find the Mean of the following sets of data.

(1935) 42, 13, 88, 19, 78, 76, 27, 59, 17

Mean = _____

(1936) 25, 41, 70, 46, 33, 12, 72

Mean = _____

(1937) 84, 81, 8, 22, 36, 58, 56, 24

Mean = _____

(1938) 6, 42, 16, 81, 84, 83, 17, 82, 37

Mean = _____

(1939) 77, 37, 54, 68, 28, 95, 44, 23

Mean = _____

(1940) 53, 63, 17, 24, 91, 8, 2, 14

Mean = _____

Measure of Center - Median

Find the Median of the following sets of data.

1941   49, 6, 57, 77, 24, 50, 84, 57

Median = ____

1942   64, 14, 36, 60, 45, 74

Median = ____

1943   34, 9, 59, 33, 69, 96, 57, 4, 56

Median = ____

1944   86, 63, 99, 83, 94, 40, 1

Median = ____

1945   14, 35, 15, 68, 36, 77, 59, 5, 14

Median = ____

1946   27, 52, 97, 4, 15, 59, 84

Median = ____

Measure of Center - Median

Find the Median of the following sets of data.

1947   78, 79, 29, 8, 64, 59, 39

Median = ___

1948   6, 56, 45, 49, 71, 14, 24

Median = ___

1949   67, 89, 17, 64, 99, 2, 46, 77

Median = ___

1950   38, 57, 50, 55, 19, 43, 44, 60, 1

Median = ___

1951   42, 28, 79, 22, 4, 66, 32, 36

Median = ___

1952   54, 96, 9, 13, 91, 28, 88, 90, 89

Median = ___

Measure of Center - Mode

Find the Mode of the following sets of data.

(1953) 59, 63, 80, 40, 15, 57

Mode = ____

(1954) 38, 2, 41, 72, 58, 10

Mode = ____

(1955) 41, 96, 62, 95, 45, 8

Mode = ____

(1956) 43, 35, 3, 64, 24, 85, 29, 45, 5

Mode = ____

(1957) 37, 64, 47, 29, 62, 64, 29, 83

Mode = _____

(1958) 80, 80, 61, 27, 70, 41, 20

Mode = ____

Measure of Center - Mode

Find the Mode of the following sets of data.

(1959) 63, 93, 74, 26, 14, 1, 47

Mode = _____

(1960) 17, 46, 20, 45, 56, 87, 94

Mode = _____

(1961) 40, 95, 11, 70, 53, 1, 37, 93, 44

Mode = _____

(1962) 96, 46, 12, 13, 81, 16, 84, 81, 73

Mode = _____

(1963) 93, 82, 2, 30, 12, 3, 32

Mode = _____

(1964) 5, 90, 48, 1, 9, 17

Mode = _____

Measure of Variability - Range

Find the Range of the following sets of data.

1965   8, 42, 89, 41, 85, 50, 57

Range = ____

1966   18, 41, 83, 69, 50, 77, 80, 2, 88

Range = ____

1967   65, 1, 62, 39, 72, 63, 5, 93

Range = ____

1968   11, 84, 9, 38, 48, 11

Range = ____

1969   88, 92, 35, 79, 39, 70

Range = ____

1970   23, 41, 31, 35, 25, 55, 53, 99

Range = ____

Measure of Variability - Range

Find the Range of the following sets of data.

1971   32, 79, 34, 23, 30, 16, 25

Range = ____

1972   47, 39, 70, 45, 45, 38, 96, 30

Range = ____

1973   1, 41, 19, 69, 31, 95, 76

Range = ____

1974   16, 98, 42, 32, 65, 25, 70, 28, 14

Range = ____

1975   17, 11, 6, 61, 31, 50, 86, 8, 32

Range = ____

1976   85, 41, 54, 65, 34, 93, 33, 27, 11

Range = ____

ANSWERS

Page 1: Multiplication with Whole Numbers

1. **1**   2. **4 3/8**   3. **2 2/3**   4. **3 3/5**   5. **1/3**   6. **2 4/5**   7. **1 1/2**

8. **1 1/2**   9. **1 1/3**   10. **5/6**   11. **2**   12. **1/2**   13. **1 1/3**   14. **4 2/3**

15. **1 1/2**   16. **1 1/4**   17. **2 1/4**   18. **3**   19. **1**   20. **1/6**

Page 2: Multiplication with Whole Numbers

21. **3 3/4**   22. **2 2/3**   23. **5/6**   24. **1 1/2**   25. **1 4/5**   26. **3 1/3**

27. **3/4**   28. **6 3/4**   29. **3**   30. **4/5**   31. **1 1/3**   32. **1 3/4**

33. **1/2**   34. **1**   35. **1 7/8**   36. **2/3**   37. **2 1/2**   38. **4 4/5**

39. **2/3**   40. **1**

Page 3: Multiplication with Whole Numbers

41. **1/8**   42. **1/2**   43. **2 2/3**   44. **4/5**   45. **1**   46. **2/5**

47. **3**   48. **4 3/8**   49. **1 1/3**   50. **4**   51. **1 7/8**   52. **1 1/3**

53. **1**   54. **1 3/5**   55. **1 1/3**   56. **4/5**   57. **1/4**   58. **3 1/3**

59. **1/3**   60. **3/4**

Page 4: Division with Whole Numbers

61. **1/16**   62. **1/6**   63. **1/15**   64. **2/27**   65. **1/40**   66. **1/6**   67. **3/40**

68. **1/36**   69. **1/28**   70. **2/9**   71. **1/5**   72. **1/12**   73. **1/12**   74. **1/12**

75. **1/14**   76. **2/3**   77. **1/8**   78. **1/20**   79. **1/24**   80. **5/56**

Page 5: Division with Whole Numbers

81. **2/9**   82. **5/56**   83. **1/30**   84. **1/8**   85. **2/15**   86. **2/21**

87. **1/9**   88. **4/15**   89. **1/12**   90. **1/16**   91. **1/35**   92. **2/3**

93. **1/18**   94. **3/32**   95. **3/8**   96. **1/10**   97. **1/8**   98. **1/6**

99. **1/4**   100. **2/27**

Page 6: Division with Whole Numbers

101. **1/6**   102. **1/4**   103. **1/32**   104. **1/25**   105. **1/6**   106. **1/24**

107. **1/12**   108. **1/54**   109. **1/9**   110. **1/35**   111. **5/48**   112. **3/8**

113. **1/12**   114. **1/4**   115. **3/5**   116. **1/48**   117. **1/24**   118. **1/32**

119. **1/54**   120. **3/5**

Page 7:   Mixed Fractions - Multiplication

121. **4 185/234**   122. **24 3/5**   123. **27 551/960**   124. **32 28/99**

125. **36 5/6**   126. **25 37/45**   127. **32 232/425**   128. **44 149/350**

129. **11 11/14**   130. **56 95/144**

Page 8:   Mixed Fractions - Multiplication

131. **36 28/75**   132. **13 7/8**   133. **11 55/72**   134. **28 1/6**

135. **19 37/92**   136. **13 31/120**   137. **8 2/3**   138. **21 16/19**

139. **20 22/25**   140. **8 207/245**

Page 9:   Mixed Fractions - Multiplication

141. **11 71/80**   142. **41 19/24**   143. **48 1471/1875**

144. **17 11/20**   145. **25 35/36**   146. **21 57/77**

147. **25 267/323**   148. **7 1/2**   149. **26 23/32**

150. **8 2/13**

Page 10:   Mixed Fractions- Division

151. **1 406/989**   152. **287/1200**   153. **1 44/45**   154. **2 130/161**

155. **3 69/265**   156. **176/219**   157. **1 377/525**   158. **7/17**

159. **804/1159**   160. **1 379/425**

Page 11:   Mixed Fractions- Division

161. **119/242**   162. **1 253/335**   163. **640/1349**

164. **1 173/640**   165. **39/146**   166. **309/1220**

167. **2 94/175**   168. **1 1129/1496**   169. **11/26**

170. **1 1/75**

Page 12: **Mixed Fractions- Division**

171. **215/423**   172. **3/5**   173. **1 311/3000**   174. **4 223/738**

175. **5 9/35**   176. **32/117**   177. **1 7/177**   178. **160/459**

179. **8/35**   180. **1 25/91**

Page 13: **Fractions: Multiple Operations**

181. **13/20**   182. **17/64**   183. **5/6**   184. **2/5**   185. **4/27**   186. **0**

187. **1 7/8**   188. **1 3/5**   189. **0**   190. **17/32**

Page 14: **Fractions: Multiple Operations**

191. **8 5/12**   192. **3/8**   193. **4**   194. **1 1/2**   195. **7/9**

196. **1 1/15**   197. **1 3/25**   198. **1 2/3**   199. **1 3/8**   200. **1 1/6**

Page 15: **Fractions: Multiple Operations**

201. **1/3**   202. **4/7**   203. **2**   204. **1 7/48**   205. **1/3**

206. **3 5/6**   207. **1 1/8**   208. **1**   209. **1/216**   210. **12**

Page 16: **Fractions: Multiple Operations**

211. **2 2/15**   212. **59/64**   213. **7/8**   214. **1 5/6**   215. **5/24**

216. **3/10**   217. **6**   218. **16/125**   219. **5 1/12**   220. **2/3**

Page 17: **Fractions: Multiple Operations**

221. **13/16**   222. **6 1/40**   223. **1/216**   224. **7/8**   225. **1 1/3**

226. **8 1/6**   227. **18/25**   228. **1 3/4**   229. **7/8**   230. **2 7/10**

Page 18: **Fractions: Multiple Operations**

231. **9/20**   232. **2/27**   233. **23/64**   234. **4/5**   235. **1 1/4**

236. **2**   237. **7/9**   238. **6 2/15**   239. **1 1/4**   240. **1/3**

Page 19: **Fractions: Multiple Operations**

241. **1/2**   242. **5/18**   243. **8/27**   244. **33/40**   245. **1/4**   246. **16/25**

247. **6**   248. **1/3**   249. **7/50**   250. **2/5**

Page 20: **Fractions: Multiple Operations**

251. **17/40**   252. **1**   253. **1/4**   254. **17/50**   255. **23/64**

256. **2/27**   257. **1 1/6**   258. **1 3/8**   259. **1 1/3**   260. **105/512**

Page 21:   Simplifying Fractions

261. **1/6**   262. **6 2/3**   263. **3**   264. **8 1/10**   265. **1/2**

266. **6**   267. **5/12**   268. **1/2**   269. **8**   270. **1/3**

271. **8 1/2**   272. **2 1/3**   273. **8**   274. **7 1/2**   275. **7**

276. **1/3**   277. **8**   278. **4**   279. **9 1/2**   280. **3**

Page 22:   Simplifying Fractions

281. **8**   282. **3/8**   283. **1/7**   284. **5**   285. **7/15**   286. **5**

287. **4**   288. **6 4/7**   289. **2**   290. **2/5**   291. **4 1/4**   292. **2/3**

293. **7/10**   294. **3/8**   295. **1/4**   296. **7 1/3**   297. **5/14**   298. **2/3**

299. **1/4**   300. **8**

Page 23:   Simplifying Fractions

301. **6 1/3**   302. **6**   303. **7**   304. **4/5**   305. **3**

306. **7 1/3**   307. **4 5/14**   308. **6**   309. **1/2**   310. **5 1/2**

311. **1/3**   312. **9 1/3**   313. **4 1/14**   314. **3 1/4**   315. **4/5**

316. **11/15**   317. **8 5/12**   318. **5 4/5**   319. **8**   320. **3/5**

Page 24:   Simplifying Fractions

321. **8**   322. **6**   323. **2**   324. **6**   325. **4/5**

326. **1/6**   327. **3**   328. **6 2/3**   329. **7/8**   330. **1/5**

331. **1/2**   332. **4**   333. **1/3**   334. **8**   335. **3**

336. **8 9/14**   337. **6**   338. **1/2**   339. **7**   340. **4/5**

Page 25:   Simplifying Fractions

341. **8 5/6**   342. **7/8**   343. **5**   344. **3**   345. **7 1/3**

346. **5 3/8**   347. **8 7/15**   348. **9 1/2**   349. **2/5**   350. **7 5/14**

351. **5**   352. **5**   353. **7**   354. **1/3**   355. **1/2**

356. **2/3**   357. **1/2**   358. **8**   359. **2 1/4**   360. **5/6**

Page 26: Simplifying Fractions

361. **6**
362. **3/10**
363. **1/3**
364. **1/2**
365. **8**

366. **8**
367. **3**
368. **8/15**
369. **1/3**
370. **1/2**

371. **3/4**
372. **6**
373. **2**
374. **2**
375. **4 1/4**

376. **8 5/14**
377. **3 1/5**
378. **8 1/5**
379. **1/3**
380. **2/3**

Page 27: Simplifying Fractions

381. **3**
382. **3/5**
383. **6 3/4**
384. **3**
385. **7**
386. **3 1/5**

387. **5/6**
388. **8 1/2**
389. **3/8**
390. **1/4**
391. **5/14**
392. **5/6**

393. **7 1/8**
394. **5 3/5**
395. **2 1/3**
396. **7 1/4**
397. **7**
398. **5 1/6**

399. **14/15**
400. **4 3/4**

Page 28: Percent

401. **130.35**
402. **177.2**
403. **1.5**
404. **312**
405. **25%**

406. **42**
407. **2%**
408. **15%**
409. **836**
410. **136.8**

411. **25%**
412. **276**
413. **16.8**
414. **10%**
415. **96.25**

416. **10%**
417. **497**
418. **8%**
419. **15%**
420. **5%**

Page 29: Percent

421. **15%**
422. **10%**
423. **520**
424. **20.55**
425. **151**
426. **25%**

427. **2%**
428. **147**
429. **11.28**
430. **69.6**
431. **15%**
432. **736**

433. **68.8**
434. **516**
435. **836**
436. **29.4**
437. **743**
438. **125.5**

439. **19.9**
440. **47**

Page 30: Percent

441. **5%**
442. **111.45**
443. **5.12**
444. **346**
445. **883**

446. **435**
447. **118**
448. **20%**
449. **625**
450. **35**

451. **8%**
452. **55.95**
453. **49.65**
454. **124.4**
455. **354**

456. **553**
457. **680**
458. **25%**
459. **10%**
460. **64.95**

Page 31: Percent

461. **11.95**   462. **143**   463. **25%**   464. **657**   465. **246**   466. **324**

467. **4**   468. **760**   469. **326**   470. **5%**   471. **8.64**   472. **30.2**

473. **5%**   474. **540**   475. **20%**   476. **572**   477. **607**   478. **65.55**

479. **84**   480. **70.3**

Page 32: Percent

481. **448**   482. **10.5**   483. **25%**   484. **2.88**   485. **63.2**

486. **15%**   487. **19.48**   488. **198.2**   489. **12.3**   490. **47.36**

491. **166**   492. **15%**   493. **37.12**   494. **5%**   495. **20%**

496. **209**   497. **597**   498. **2%**   499. **239.25**   500. **41.9**

Page 33: Percent and Decimals

501. **0.85**   502. **0.29**   503. **0.52**   504. **0.16**   505. **0.25**   506. **0.56**

507. **0.53**   508. **0.44**   509. **0.84**   510. **0.07**   511. **0.62**   512. **0.32**

513. **0.82**   514. **0.35**   515. **0.34**   516. **0.47**   517. **0.75**   518. **0.54**

519. **0.72**   520. **0.93**

Page 34: Percent and Decimals

521. **0.21**   522. **0.11**   523. **0.57**   524. **0.97**   525. **0.3**   526. **0.82**

527. **0.67**   528. **0.98**   529. **0.29**   530. **0.39**   531. **0.1**   532. **0.61**

533. **0.13**   534. **0.68**   535. **0.38**   536. **0.09**   537. **0.76**   538. **0.79**

539. **0.75**   540. **0.37**

Page 35: Percent and Decimals

541. **0.25**   542. **0.73**   543. **0.7**   544. **0.8**   545. **0.46**   546. **0.66**

547. **0.78**   548. **0.93**   549. **0.11**   550. **0.18**   551. **0.42**   552. **0.69**

553. **0.33**   554. **0.72**   555. **0.92**   556. **0.32**   557. **0.79**   558. **0.88**

559. **0.54**   560. **0.59**

Page 36: Percent and Decimals

561. **0.79** 562. **0.9** 563. **0.5** 564. **0.36** 565. **0.06** 566. **0.82**

567. **0.66** 568. **0.93** 569. **0.95** 570. **0.2** 571. **0.21** 572. **0.77**

573. **1** 574. **0.52** 575. **0.69** 576. **0.71** 577. **0.43** 578. **0.44**

579. **0.98** 580. **0.88**

Page 37: Percent and Decimals

581. **20%** 582. **54%** 583. **22%** 584. **91%** 585. **64%** 586. **96%** 587. **39%**

588. **95%** 589. **49%** 590. **44%** 591. **50%** 592. **84%** 593. **47%** 594. **41%**

595. **85%** 596. **12%** 597. **1%** 598. **17%** 599. **99%** 600. **21%**

Page 38: Percent and Decimals

601. **65%** 602. **70%** 603. **33%** 604. **20%** 605. **24%** 606. **61%** 607. **36%**

608. **44%** 609. **56%** 610. **72%** 611. **59%** 612. **29%** 613. **46%** 614. **42%**

615. **87%** 616. **28%** 617. **99%** 618. **47%** 619. **81%** 620. **78%**

Page 39: Percent and Decimals

621. **28%** 622. **16%** 623. **53%** 624. **69%** 625. **49%** 626. **27%**

627. **3%** 628. **60%** 629. **87%** 630. **4%** 631. **56%** 632. **2%**

633. **94%** 634. **63%** 635. **19%** 636. **71%** 637. **88%** 638. **100%**

639. **95%** 640. **12%**

Page 40: Percent and Decimals

641. **88%** 642. **48%** 643. **92%** 644. **99%** 645. **84%** 646. **3%** 647. **70%**

648. **18%** 649. **43%** 650. **38%** 651. **55%** 652. **80%** 653. **52%** 654. **75%**

655. **34%** 656. **33%** 657. **85%** 658. **69%** 659. **86%** 660. **56%**

Page 41: Percent - Advanced

661. **1.7%** 662. **4.0%** 663. **3.7%** 664. **633** 665. **277**

666. **29.412** 667. **0.4%** 668. **26.042** 669. **607** 670. **6**

671. **5.0%** 672. **0.328** 673. **0.4%** 674. **7.8%** 675. **9**

676. 1.34   677. 0.2%   678. 0.435   679. 36.312   680. 0.171

Page 42:   Percent - Advanced
681. 0.63   682. 25   683. 32.439   684. 3.7%   685. 22.83
686. 613   687. 6   688. 951   689. 7.3%   690. 6
691. 0.2%   692. 0.042   693. 7.9%   694. 0.5%   695. 40
696. 17.996   697. 0.444   698. 11.136   699. 960   700. 3.4%

Page 43:   Percent - Advanced
701. 17.442   702. 251   703. 6.1%   704. 0.124   705. 3.0%
706. 977   707. 63   708. 0.57   709. 0.7%   710. 8.8%
711. 59   712. 0.044   713. 7.65   714. 263   715. 3.5%
716. 45.771   717. 0.5%   718. 0.6%   719. 0.085   720. 0.335

Page 44:   Percent - Advanced
721. 0.096   722. 0.075   723. 4.3%   724. 11   725. 0.666
726. 45   727. 0.5%   728. 18.833   729. 0.007   730. 0.2%
731. 32.64   732. 4   733. 626   734. 0.005   735. 0.2%
736. 6   737. 784   738. 616   739. 506   740. 3.3%

Page 45:   Percent - Advanced
741. 0.792   742. 2   743. 4.5%   744. 3.1%   745. 858   746. 7
747. 49   748. 98   749. 103   750. 4.525 751. 6.5%   752. 167
753. 0.2%   754. 0.536   755. 472   756. 1.4%   757. 1.98   758. 2.312
759. 8.9%   760. 2

Page 46:   Percent - Advanced
761. 491   762. 0.7   763. 7.0%   764. 66   765. 5.9%   766. 0.096
767. 0.1%   768. 2.012   769. 0.3%   770. 1.218 771. 699   772. 0.2%
773. 3.8%   774. 966   775. 1.6%   776. 2.976 777. 6.3%   778. 6.4%
779. 0.006   780. 1.025

Page 47: Percent - Advanced

781. **0.6%** 782. **43.035** 783. **0.2%** 784. **5.4%** 785. **2.5%**

786. **79** 787. **7** 788. **0.6%** 789. **2** 790. **13.16**

791. **475** 792. **2.9%** 793. **7.6** 794. **2** 795. **0.3%**

796. **1.715** 797. **0.264** 798. **0.201** 799. **827** 800. **5**

Page 48: Percent - Advanced

801. **7** 802. **4** 803. **0.004** 804. **4.6%** 805. **974**

806. **9.8%** 807. **30** 808. **0.656** 809. **52** 810. **89**

811. **3** 812. **0.1%** 813. **19** 814. **40** 815. **2**

816. **0.075** 817. **20.414** 818. **932** 819. **47** 820. **0.364**

Page 49: Ratio Conversions

821.

	Ratio	Fraction	Percent	Decimal
a.	2:2	2/2	100%	1
b.	4:6	4/6	66.7%	0.667
c.	4:5	4/5	80%	0.8
d.	1:8	1/8	12.5%	0.125
e.	3:6	3/6	50%	0.5
f.	2:8	2/8	25%	0.25
g.	2:3	2/3	66.7%	0.667
h.	1:9	1/9	11.1%	0.111
i.	3:9	3/9	33.3%	0.333
j.	5:6	5/6	83.3%	0.833
k.	5:7	5/7	71.4%	0.714
l.	1:3	1/3	33.3%	0.333
m.	7:9	7/9	77.8%	0.778

Page 50: Ratio Conversions

822.

	Ratio	Fraction	Percent	Decimal
a.	2:10	2/10	20%	0.2
b.	6:7	6/7	85.7%	0.857
c.	7:8	7/8	87.5%	0.875
d.	2:3	2/3	66.7%	0.667
e.	5:5	5/5	100%	1
f.	1:6	1/6	16.7%	0.167
g.	1:7	1/7	14.3%	0.143
h.	3:7	3/7	42.9%	0.429
i.	4:8	4/8	50%	0.5
j.	9:10	9/10	90%	0.9
k.	2:8	2/8	25%	0.25
l.	6:8	6/8	75%	0.75
m.	1:2	1/2	50%	0.5

Page 51:   Ratio Conversions

823.

	Ratio	Fraction	Percent	Decimal
a.	2:4	2/4	50%	0.5
b.	1:2	1/2	50%	0.5
c.	4:6	4/6	66.7%	0.667
d.	2:10	2/10	20%	0.2
e.	8:10	8/10	80%	0.8
f.	6:8	6/8	75%	0.75
g.	3:3	3/3	100%	1
h.	2:7	2/7	28.6%	0.286
i.	1:3	1/3	33.3%	0.333
j.	5:7	5/7	71.4%	0.714
k.	4:9	4/9	44.4%	0.444
l.	3:8	3/8	37.5%	0.375
m.	3:5	3/5	60%	0.6

Page 52:   Ratio Conversions

824.

	Ratio	Fraction	Percent	Decimal
a.	2:4	2/4	50%	0.5
b.	1:1	1/1	100%	1
c.	1:7	1/7	14.3%	0.143
d.	2:3	2/3	66.7%	0.667
e.	3:5	3/5	60%	0.6
f.	1:2	1/2	50%	0.5
g.	2:9	2/9	22.2%	0.222
h.	4:7	4/7	57.1%	0.571
i.	1:3	1/3	33.3%	0.333
j.	5:7	5/7	71.4%	0.714
k.	3:4	3/4	75%	0.75
l.	1:8	1/8	12.5%	0.125
m.	2:6	2/6	33.3%	0.333

Page 53:   Cartesian Coordinates

825.

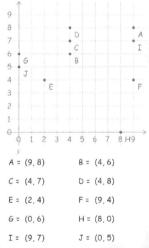

A = (9, 8)   B = (4, 6)

C = (4, 7)   D = (4, 8)

E = (2, 4)   F = (9, 4)

G = (0, 6)   H = (8, 0)

I = (9, 7)   J = (0, 5)

Page 54: Cartesian Coordinates

826.

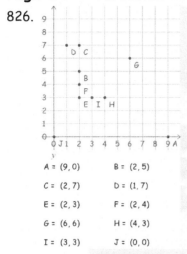

A = (9, 0)   B = (2, 5)

C = (2, 7)   D = (1, 7)

E = (2, 3)   F = (2, 4)

G = (6, 6)   H = (4, 3)

I = (3, 3)   J = (0, 0)

Page 55: Cartesian Coordinates

827.

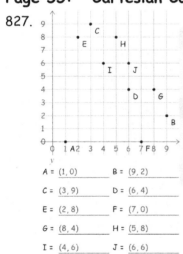

A = (1, 0)   B = (9, 2)

C = (3, 9)   D = (6, 4)

E = (2, 8)   F = (7, 0)

G = (8, 4)   H = (5, 8)

I = (4, 6)   J = (6, 6)

Page 56:   Cartesian Coordinates

828.

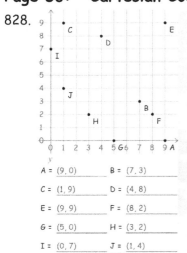

A = (9, 0)   B = (7, 3)

C = (1, 9)   D = (4, 8)

E = (9, 9)   F = (8, 2)

G = (5, 0)   H = (3, 2)

I = (0, 7)   J = (1, 4)

Page 57:   Cartesian Coordinates

829.

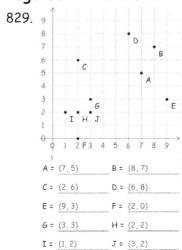

A = (7, 5)   B = (8, 7)

C = (2, 6)   D = (6, 8)

E = (9, 3)   F = (2, 0)

G = (3, 3)   H = (2, 2)

I = (1, 2)   J = (3, 2)

Page 58: Cartesian Coordinates With Four Quadrants

830.

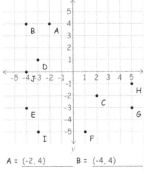

A = (-2, 4)　　B = (-4, 4)

C = (2, -2)　　D = (-3, 1)

E = (-4, -3)　　F = (1, -5)

G = (5, -3)　　H = (5, -1)

I = (-3, -5)　　J = (-4, 0)

Page 59: Cartesian Coordinates With Four Quadrants

831.

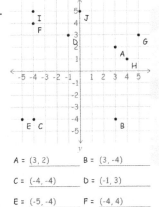

A = (3, 2)　　B = (3, -4)

C = (-4, -4)　　D = (-1, 3)

E = (-5, -4)　　F = (-4, 4)

G = (5, 3)　　H = (4, 1)

I = (-4, 5)　　J = (0, 5)

Page 60: Cartesian Coordinates With Four Quadrants

832.

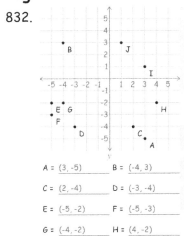

A = (3, -5)	B = (-4, 3)
C = (2, -4)	D = (-3, -4)
E = (-5, -2)	F = (-5, -3)
G = (-4, -2)	H = (4, -2)
I = (3, 1)	J = (1, 3)

Page 61: Cartesian Coordinates With Four Quadrants

833.

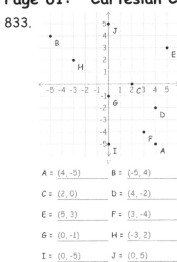

A = (4, -5)	B = (-5, 4)
C = (2, 0)	D = (4, -2)
E = (5, 3)	F = (3, -4)
G = (0, -1)	H = (-3, 2)
I = (0, -5)	J = (0, 5)

Page 62: Cartesian Coordinates With Four Quadrants

834.

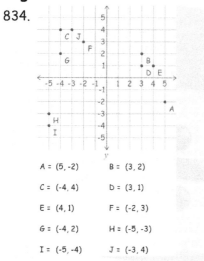

A = (5, -2)	B = (3, 2)
C = (-4, 4)	D = (3, 1)
E = (4, 1)	F = (-2, 3)
G = (-4, 2)	H = (-5, -3)
I = (-5, -4)	J = (-3, 4)

Page 63: Cartesian Coordinates With Four Quadrants

835.

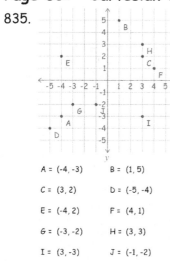

A = (-4, -3)	B = (1, 5)
C = (3, 2)	D = (-5, -4)
E = (-4, 2)	F = (4, 1)
G = (-3, -2)	H = (3, 3)
I = (3, -3)	J = (-1, -2)

Page 64: Cartesian Coordinates With Four Quadrants

836.

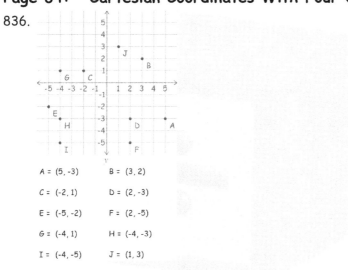

A = (5, -3)   B = (3, 2)

C = (-2, 1)   D = (2, -3)

E = (-5, -2)   F = (2, -5)

G = (-4, 1)   H = (-4, -3)

I = (-4, -5)   J = (1, 3)

Page 65: Cartesian Coordinates With Four Quadrants

837.

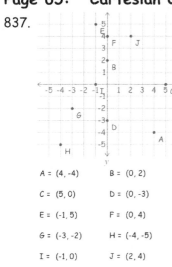

A = (4, -4)   B = (0, 2)

C = (5, 0)   D = (0, -3)

E = (-1, 5)   F = (0, 4)

G = (-3, -2)   H = (-4, -5)

I = (-1, 0)   J = (2, 4)

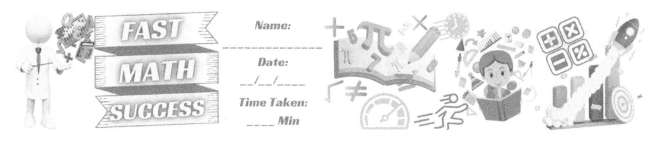

Page 66:   Cartesian Coordinates With Four Quadrants

838.

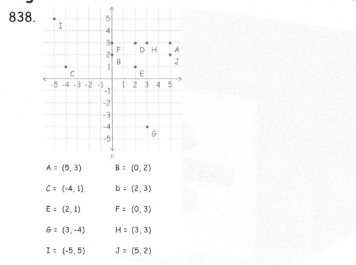

A = (5, 3)	B = (0, 2)
C = (-4, 1)	D = (2, 3)
E = (2, 1)	F = (0, 3)
G = (3, -4)	H = (3, 3)
I = (-5, 5)	J = (5, 2)

Page 67:   Plot Lines

839.

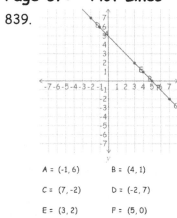

A = (-1, 6)	B = (4, 1)
C = (7, -2)	D = (-2, 7)
E = (3, 2)	F = (5, 0)

Page 68:   Plot Lines

840.

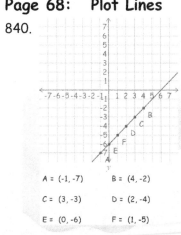

A = (-1, -7)	B = (4, -2)
C = (3, -3)	D = (2, -4)
E = (0, -6)	F = (1, -5)

Page 69:   Plot Lines

841.

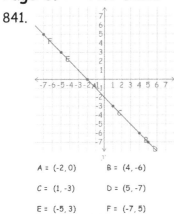

A = (-2, 0)   B = (4, -6)

C = (1, -3)   D = (5, -7)

E = (-5, 3)   F = (-7, 5)

Page 70:   Plot Lines

842.

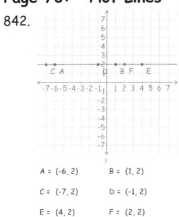

A = (-6, 2)   B = (1, 2)

C = (-7, 2)   D = (-1, 2)

E = (4, 2)   F = (2, 2)

Page 71:   Plot Lines

843.

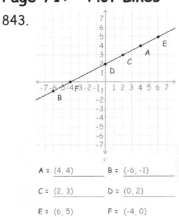

A = (4, 4)   B = (-6, -1)

C = (2, 3)   D = (0, 2)

E = (6, 5)   F = (-4, 0)

Name:

Date:
__/__/____
Time Taken:
____ Min

Page 72:   Plot Lines

844.
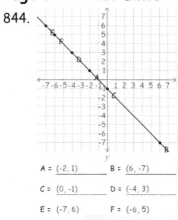

A = (-2, 1)   B = (6, -7)

C = (0, -1)   D = (-4, 3)

E = (-7, 6)   F = (-6, 5)

Page 73:   Plot Lines

845.
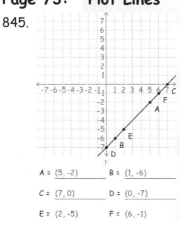

A = (5, -2)   B = (1, -6)

C = (7, 0)   D = (0, -7)

E = (2, -5)   F = (6, -1)

Page 74:   Plot Lines

846.
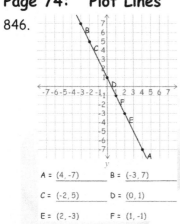

A = (4, -7)   B = (-3, 7)

C = (-2, 5)   D = (0, 1)

E = (2, -3)   F = (1, -1)

Page 75:   Exponents

847. **2,809**   848. **8,000**   849. **1,600**   850. **373,248**   851. **125,000**

Fast Math Success Workbook Grade 6-7

852. 148,877   853. 704,969   854. 857,375   855. 7,921   856. 7,744

857. 300,763   858. 5,625   859. 140,608   860. 8,649   861. 1,728

862. 12,167   863. 970,299   864. 7,225   865. 287,496   866. 166,375

Page 76:   Exponents

867. 54,872   868. 4,913   869. 1,849   870. 625   871. 6,241

872. 3,025   873. 343   874. 658,503   875. 9,604   876. 9,025

877. 2,704   878. 778,688   879. 9   880. 9,261   881. 85,184

882. 16   883. 8,000   884. 175,616   885. 857,375   886. 68,921

Page 77:   Exponents

887. 3,844   888. 185,193   889. 328,509   890. 216   891. 6,889

892. 117,649   893. 592,704   894. 531,441   895. 4,900   896. 27,000

897. 175,616   898. 1   899. 6,561   900. 2,601   901. 4,913

902. 8,281   903. 140,608   904. 1,728   905. 373,248   906. 456,533

Page 78:   Exponents

907. 3,600   908. 9,801   909. 68,921   910. 405,224   911. 140,608

912. 841   913. 2,025   914. 4,900   915. 9,025   916. 7,396

917. 941,192   918. 314,432   919. 804,357   920. 27,000   921. 3,025

922. 778,688   923. 166,375   924. 32,768   925. 9,604   926. 970,299

Page 79:   Scientific Notation

927. 2.35×10^6 928. 2.5×10^6 929. 2.56×10^5

930. 8.7×10^6 931. 9.21×10^6 932. 4.1×10^6

933. 4.39×10^6 934. 8.51×10^6 935. 9.5×10^6

936. 3.5×10^6 937. 2.6×10^6 938. 9.8×10^6

939. 1.084×10^6 940. 4.18×10^6 941. 8.735×10^6

942. 5.62×10^6 943. 3.53×10^6 944. 7.625×10^6

945. 4.936×10^6 946. 2.577×10^6

Page 80: Scientific Notation

947. **8.32 × 10^6**   948. **9.4 × 10^6**   949. **3.25 × 10^6**

950. **4.956 × 10^6**   951. **6.94 × 10^6**   952. **6 × 10^6**

953. **6.286 × 10^6**   954. **7.355 × 10^6**   955. **6.6 × 10^6**

956. **5.4 × 10^6**   957. **1.65 × 10^6**   958. **6.8 × 10^6**

959. **7.891 × 10^6**   960. **8.29 × 10^6**   961. **7 × 10^6**

962. **9.5 × 10^5**   963. **8.34 × 10^6**   964. **1.883 × 10^6**

965. **3.33 × 10^6**   966. **7.81 × 10^6**

Page 81: Scientific Notation

967. **4.35 × 10^6**   968. **7.16 × 10^6**   969. **1.89 × 10^6**

970. **9.51 × 10^6**   971. **8.708 × 10^6**   972. **8.7 × 10^5**

973. **3.2 × 10^6**   974. **7.03 × 10^6**   975. **6.993 × 10^6**

976. **8.44 × 10^6**   977. **6.71 × 10^6**   978. **1.27 × 10^6**

979. **2.37 × 10^6**   980. **3.8 × 10^6**   981. **5.36 × 10^6**

982. **6.2 × 10^6**   983. **3.622 × 10^6**   984. **6.59 × 10^6**

985. **7.4 × 10^6**   986. **9.4 × 10^6**

Page 82: Scientific Notation

987. **8,800,000**   988. **9,624,000**   989. **86,000**   990. **7,570,000**

991. **7,825,000**   992. **1,750,000**   993. **5,000,000**   994. **9,357,000**

995. **4,600,000**   996. **3,025,000**   997. **9,536,000**   998. **920,000**

999. **1,727,000**   1000. **9,870,000**   1001. **6,300,000**   1002. **8,700,000**

1003. **6,040,000**   1004. **6,900,000**   1005. **5,900,000**   1006. **7,652,000**

Page 83: Scientific Notation

1007. **3,603,000**   1008. **6,800,000**   1009. **8,100,000**   1010. **22,000**

1011. **1,130,000**   1012. **8,400,000**   1013. **4,900,000**   1014. **1,075,000**

1015. **7,200,000**   1016. **7,570,000**   1017. **1,680,000**   1018. **1,840,000**

1019. **6,500,000**   1020. **6,870,000**   1021. **730,000**   1022. **5,262,000**

1023. **9,000,000**   1024. **4,470,000**   1025. **3,200,000**   1026. **9,481,000**

Page 84:   Scientific Notation

1027. **3,830,000**   1028. **140,000**   1029. **7,030,000**   1030. **9,000,000**

1031. **3,639,000**   1032. **4,970,000**   1033. **7,330,000**   1034. **5,464,000**

1035. **5,300,000**   1036. **4,543,000**   1037. **8,134,000**   1038. **990,000**

1039. **3,700,000**   1040. **6,289,000**   1041. **6,260,000**   1042. **2,944,000**

1043. **9,870,000**   1044. **6,400,000**   1045. **9,200,000**   1046. **2,000,000**

Page 85:   Expressions - Single Step

1047. **7**   1048. **4**   1049. **7**   1050. **1**   1051. **4**   1052. **8**   1053. **2**   1054. **4**   1055. **4**

1056. **9**   1057. **6**   1058. **7**   1059. **6**   1060. **5**   1061. **3**   1062. **7**   1063. **1**   1064. **6**

1065. **2**   1066. **3**

Page 86:   Expressions - Single Step

1067. **1**   1068. **3**   1069. **4**   1070. **7**   1071. **1**   1072. **8**   1073. **1**   1074. **7**   1075. **5**

1076. **1**   1077. **1**   1078. **7**   1079. **8**   1080. **2**   1081. **3**   1082. **2**   1083. **4**   1084. **4**

1085. **7**   1086. **3**

Page 87:   Expressions - Single Step

1087. **4**   1088. **6**   1089. **1**   1090. **3**   1091. **1**   1092. **2**   1093. **7**   1094. **6**   1095. **4**

1096. **7**   1097. **9**   1098. **6**   1099. **6**   1100. **3**   1101. **6**   1102. **8**   1103. **4**   1104. **8**

1105. **8**   1106. **7**

Page 88:   Expressions - Single Step

1107. **5**   1108. **9**   1109. **7**   1110. **5**   1111. **8**   1112. **2**   1113. **9**   1114. **2**   1115. **2**

1116. **8**   1117. **3**   1118. **5**   1119. **9**   1120. **1**   1121. **4**   1122. **1**   1123. **4**   1124. **6**

1125. **2**   1126. **3**

Page 89:   Expressions - Single Step

1127. **8**   1128. **8**   1129. **1**   1130. **2**   1131. **4**   1132. **4**   1133. **3**   1134. **2**   1135. **7**

1136. **4**   1137. **1**   1138. **1**   1139. **5**   1140. **1**   1141. **6**   1142. **2**   1143. **7**   1144. **4**

1145. **1**   1146. **3**

Page 90:   Number Problems

1147. **24**   1148. **25**   1149. **16**   1150. **26**   1151. **19**   1152. **54**   1153. **30**   1154. **11**

1155. **38**   1156. **6**

Page 91:   Number Problems

1157. **15**   1158. **23**   1159. **21**   1160. **24**   1161. **25**   1162. **21**   1163. **20**   1164. **34**

1165. **36**   1166. **18**

Page 92:   Number Problems

1167. **33**   1168. **12**   1169. **29**   1170. **32**   1171. **17**   1172. **27**   1173. **7**   1174. **22**

1175. **31**   1176. **24**

Page 93:   Pre-Algebra Equations (One Step) Addition and Subtraction

1177. $x = 3$ 1178. $x = 9$ 1179. $x = 6$ 1180. $x = 7$ 1181. $x = 2$ 1182. $x = 2$

1183. $x = 7$ 1184. $x = 1$ 1185. $x = 4$ 1186. $x = 2$ 1187. $x = 4$ 1188. $x = 5$

1189. $x = 6$ 1190. $x = 7$ 1191. $x = 6$ 1192. $x = 8$ 1193. $x = 3$ 1194. $x = 5$

1195. $x = 8$ 1196. $x = 3$

Page 94:   Pre-Algebra Equations (One Step) Addition and Subtraction

1197. $x = 6$ 1198. $x = 3$ 1199. $x = 6$ 1200. $x = 8$ 1201. $x = 2$

1202. $x = 6$ 1203. $x = 3$ 1204. $x = 9$ 1205. $x = 2$ 1206. $x = 7$

1207. $x = 5$ 1208. $x = 3$ 1209. $x = 5$ 1210. $x = 5$ 1211. $x = 9$

1212. $x = 9$ 1213. $x = 2$ 1214. $x = 3$ 1215. $x = 7$ 1216. $x = 9$

Page 95:   Pre-Algebra Equations (One Step) Addition and Subtraction

1217. $x = 5$ 1218. $x = 7$ 1219. $x = 6$ 1220. $x = 5$ 1221. $x = 9$

1222. $x = 4$ 1223. $x = 8$ 1224. $x = 8$ 1225. $x = 9$ 1226. $x = 3$

1227. $x = 1$ 1228. $x = 6$ 1229. $x = 5$ 1230. $x = 7$ 1231. $x = 1$

1232. $x = 9$ 1233. $x = 4$ 1234. $x = 9$ 1235. $x = 3$ 1236. $x = 6$

Page 96: Pre-Algebra Equations (One Step) Addition and Subtraction

1237. x = 2   1238. x = 9   1239. x = 3   1240. x = 3   1241. x = 5

1242. x = 1   1243. x = 6   1244. x = 8   1245. x = 5   1246. x = 4

1247. x = 3   1248. x = 8   1249. x = 5   1250. x = 5   1251. x = 4

1252. x = 2   1253. x = 2   1254. x = 6   1255. x = 6   1256. x = 7

Page 97: Pre-Algebra Equations (One Step) Addition and Subtraction

1257. x = 9   1258. x = 9   1259. x = 1   1260. x = 5   1261. x = 2

1262. x = 3   1263. x = 6   1264. x = 4   1265. x = 4   1266. x = 8

1267. x = 9   1268. x = 5   1269. x = 8   1270. x = 8   1271. x = 4

1272. x = 8   1273. x = 7   1274. x = 5   1275. x = 6   1276. x = 5

Page 98: Pre-Algebra Equations (One Step) Addition and Subtraction

1277. x = 9   1278. x = 7   1279. x = 5   1280. x = 7   1281. x = 7

1282. x = 5   1283. x = 5   1284. x = 5   1285. x = 8   1286. x = 3

1287. x = 4   1288. x = 7   1289. x = 2   1290. x = 7   1291. x = 5

1292. x = 4   1293. x = 2   1294. x = 9   1295. x = 1   1296. x = 3

Page 99: Pre-Algebra Equations (One Step) Addition and Subtraction

1297. x = 1   1298. x = 9   1299. x = 4   1300. x = 9   1301. x = 5

1302. x = 6   1303. x = 1   1304. x = 6   1305. x = 8   1306. x = 9

1307. x = 7   1308. x = 2   1309. x = 6   1310. x = 7   1311. x = 2

1312. x = 2   1313. x = 5   1314. x = 3   1315. x = 9   1316. x = 1

Page 100: Pre-Algebra Equations (One Step) Addition and Subtraction

1317. x = 3   1318. x = 9   1319. x = 6   1320. x = 9   1321. x = 5

1322. x = 8   1323. x = 9   1324. x = 5   1325. x = 3   1326. x = 5

1327. x = 6   1328. x = 2   1329. x = 2   1330. x = 4   1331. x = 5

1332. x = 1   1333. x = 4   1334. x = 9   1335. x = 3   1336. x = 4

Page 101:   Pre-Algebra Equations (One Step) Addition and Subtraction

1337. x = 8	1338. x = 3	1339. x = 6	1340. x = 4	1341. x = 9
1342. x = 2	1343. x = 2	1344. x = 6	1345. x = 4	1346. x = 2
1347. x = 4	1348. x = 2	1349. x = 5	1350. x = 1	1351. x = 9
1352. x = 3	1353. x = 9	1354. x = 6	1355. x = 5	1356. x = 7

Page 102:   Pre-Algebra Equations (One Step) Addition and Subtraction

1357. x = 2	1358. x = 2	1359. x = 1	1360. x = 7	1361. x = 5
1362. x = 4	1363. x = 5	1364. x = 6	1365. x = 4	1366. x = 7
1367. x = 7	1368. x = 2	1369. x = 7	1370. x = 8	1371. x = 3
1372. x = 5	1373. x = 1	1374. x = 4	1375. x = 5	1376. x = 7

Page 103:   Pre-Algebra Equations (One Step) Multiplication and Division

1377. x = 9	1378. x = 9	1379. x = 7	1380. x = 6	1381. x = 8
1382. x = 1	1383. x = 3	1384. x = 8	1385. x = 3	1386. x = 9
1387. x = 30	1388. x = 7	1389. x = 1	1390. x = 4	1391. x = 1
1392. x = 7	1393. x = 3	1394. x = 9	1395. x = 4	1396. x = 2

Page 104:   Pre-Algebra Equations (One Step) Multiplication and Division

1397. x = 63	1398. x = 3	1399. x = 5	1400. x = 2	1401. x = 3
1402. x = 4	1403. x = 4	1404. x = 3	1405. x = 4	1406. x = 8
1407. x = 1	1408. x = 8	1409. x = 9	1410. x = 3	1411. x = 6
1412. x = 12	1413. x = 2	1414. x = 81	1415. x = 5	1416. x = 6

Page 105:   Pre-Algebra Equations (One Step) Multiplication and Division

1417. x = 4	1418. x = 9	1419. x = 4	1420. x = 9	1421. x = 54
1422. x = 4	1423. x = 8	1424. x = 9	1425. x = 9	1426. x = 6
1427. x = 8	1428. x = 8	1429. x = 2	1430. x = 2	1431. x = 5
1432. x = 5	1433. x = 48	1434. x = 4	1435. x = 2	1436. x = 1

Name:
_____
Date:
__/__/____
Time Taken:
_____ Min

Page 106: Pre-Algebra Equations (One Step) Multiplication and Division

1437. $x = 30$ 　　1438. $x = 12$ 　　1439. $x = 9$ 　　1440. $x = 35$ 　　1441. $x = 4$

1442. $x = 1$ 　　1443. $x = 8$ 　　1444. $x = 1$ 　　1445. $x = 6$ 　　1446. $x = 8$

1447. $x = 8$ 　　1448. $x = 2$ 　　1449. $x = 7$ 　　1450. $x = 6$ 　　1451. $x = 3$

1452. $x = 2$ 　　1453. $x = 6$ 　　1454. $x = 7$ 　　1455. $x = 4$ 　　1456. $x = 7$

Page 107: Pre-Algebra Equations (One Step) Multiplication and Division

1457. $x = 9$ 　　1458. $x = 4$ 　　1459. $x = 24$ 　　1460. $x = 56$ 　　1461. $x = 14$

1462. $x = 1$ 　　1463. $x = 2$ 　　1464. $x = 3$ 　　1465. $x = 5$ 　　1466. $x = 8$

1467. $x = 1$ 　　1468. $x = 5$ 　　1469. $x = 7$ 　　1470. $x = 6$ 　　1471. $x = 40$

1472. $x = 5$ 　　1473. $x = 1$ 　　1474. $x = 4$ 　　1475. $x = 7$ 　　1476. $x = 2$

Page 108: Pre-Algebra Equations (One Step) Multiplication and Division

1477. $x = 8$ 　　1478. $x = 3$ 　　1479. $x = 6$ 　　1480. $x = 1$ 　　1481. $x = 4$

1482. $x = 1$ 　　1483. $x = 1$ 　　1484. $x = 9$ 　　1485. $x = 5$ 　　1486. $x = 2$

1487. $x = 1$ 　　1488. $x = 3$ 　　1489. $x = 8$ 　　1490. $x = 6$ 　　1491. $x = 9$

1492. $x = 6$ 　　1493. $x = 5$ 　　1494. $x = 5$ 　　1495. $x = 4$ 　　1496. $x = 9$

Page 109: Pre-Algebra Equations (One Step) Multiplication and Division

1497. $x = 2$ 　　1498. $x = 2$ 　　1499. $x = 1$ 　　1500. $x = 24$ 　　1501. $x = 6$

1502. $x = 7$ 　　1503. $x = 3$ 　　1504. $x = 12$ 　　1505. $x = 1$ 　　1506. $x = 9$

1507. $x = 6$ 　　1508. $x = 5$ 　　1509. $x = 3$ 　　1510. $x = 9$ 　　1511. $x = 9$

1512. $x = 2$ 　　1513. $x = 4$ 　　1514. $x = 1$ 　　1515. $x = 2$ 　　1516. $x = 16$

Page 110: Pre-Algebra Equations (One Step) Multiplication and Division

1517. $x = 9$ 　　1518. $x = 5$ 　　1519. $x = 9$ 　　1520. $x = 6$ 　　1521. $x = 3$

1522. $x = 8$ 　　1523. $x = 4$ 　　1524. $x = 4$ 　　1525. $x = 9$ 　　1526. $x = 2$

1527. $x = 5$ 　　1528. $x = 8$ 　　1529. $x = 21$ 　　1530. $x = 1$ 　　1531. $x = 2$

1532. $x = 8$ 　　1533. $x = 9$ 　　1534. $x = 5$ 　　1535. $x = 8$ 　　1536. $x = 6$

Fast Math Success Workbook Grade 6-7   **189**

Page 111: Pre-Algebra Equations (One Step) Multiplication and Division

1537. $x = 9$	1538. $x = 4$	1539. $x = 1$	1540. $x = 1$	1541. $x = 6$
1542. $x = 8$	1543. $x = 2$	1544. $x = 3$	1545. $x = 3$	1546. $x = 7$
1547. $x = 6$	1548. $x = 20$	1549. $x = 2$	1550. $x = 9$	1551. $x = 27$
1552. $x = 5$	1553. $x = 7$	1554. $x = 30$	1555. $x = 5$	1556. $x = 6$

Page 112: Pre-Algebra Equations (One Step) Multiplication and Division

1557. $x = 7$	1558. $x = 1$	1559. $x = 9$	1560. $x = 1$	1561. $x = 5$
1562. $x = 7$	1563. $x = 6$	1564. $x = 3$	1565. $x = 1$	1566. $x = 3$
1567. $x = 5$	1568. $x = 4$	1569. $x = 3$	1570. $x = 5$	1571. $x = 45$
1572. $x = 9$	1573. $x = 9$	1574. $x = 4$	1575. $x = 3$	1576. $x = 8$

Page 113: Pre-Algebra Equations (One Step) Multiplication and Division

1577. $x = 1$	1578. $x = 72$	1579. $x = 5$	1580. $x = 6$	1581. $x = 3$
1582. $x = 2$	1583. $x = 9$	1584. $x = 1$	1585. $x = 8$	1586. $x = 7$
1587. $x = 8$	1588. $x = 1$	1589. $x = 6$	1590. $x = 1$	1591. $x = 2$
1592. $x = 30$	1593. $x = 7$	1594. $x = 27$	1595. $x = 3$	1596. $x = 25$

Page 114: Pre-Algebra Equations (One Step) Multiplication and Division

1597. $x = 3$	1598. $x = 6$	1599. $x = 3$	1600. $x = 1$	1601. $x = 6$
1602. $x = 9$	1603. $x = 3$	1604. $x = 8$	1605. $x = 2$	1606. $x = 32$
1607. $x = 1$	1608. $x = 5$	1609. $x = 1$	1610. $x = 4$	1611. $x = 9$
1612. $x = 9$	1613. $x = 6$	1614. $x = 7$	1615. $x = 4$	1616. $x = 27$

Page 115: Pre-Algebra Equations (One Step) Multiplication and Division

1617. $x = 4$	1618. $x = 3$	1619. $x = 6$	1620. $x = 63$	1621. $x = 9$
1622. $x = 3$	1623. $x = 1$	1624. $x = 1$	1625. $x = 9$	1626. $x = 9$
1627. $x = 7$	1628. $x = 8$	1629. $x = 30$	1630. $x = 1$	1631. $x = 5$
1632. $x = 5$	1633. $x = 4$	1634. $x = 5$	1635. $x = 3$	1636. $x = 9$

Page 116:   Pre-Algebra Equations (Two Sides)

1637. x = 3   1638. x = 6   1639. x = 5   1640. x = 4   1641. x = 3

1642. x = 9   1643. x = 4   1644. x = 6   1645. x = 9   1646. x = 6

Page 117:   Pre-Algebra Equations (Two Sides)

1647. x = 7   1648. x = 8   1649. x = 6   1650. x = 6   1651. x = 8

1652. x = 5   1653. x = 5   1654. x = 6   1655. x = 2   1656. x = 6

Page 118:   Pre-Algebra Equations (Two Sides)

1657. x = 6   1658. x = 4   1659. x = 7   1660. x = 7   1661. x = 5

1662. x = 9   1663. x = 9   1664. x = 3   1665. x = 9   1666. x = 3

Page 119:   Pre-Algebra Equations (Two Sides)

1667. x = 2   1668. x = 2   1669. x = 4   1670. x = 9   1671. x = 3

1672. x = 6   1673. x = 6   1674. x = 3   1675. x = 8   1676. x = 9

Page 120:   Pre-Algebra Equations (Two Sides)

1677. x = 2   1678. x = 7   1679. x = 9   1680. x = 8   1681. x = 9

1682. x = 8   1683. x = 4   1684. x = 6   1685. x = 5   1686. x = 7

Page 121:   Pre-Algebra Equations (Two Sides)

1687. x = 7   1688. x = 3   1689. x = 3   1690. x = 9   1691. x = 4

1692. x = 5   1693. x = 6   1694. x = 9   1695. x = 2   1696. x = 9

Page 122:   Pre-Algebra Equations (Two Sides)

1697. x = 5   1698. x = 6   1699. x = 7   1700. x = 9   1701. x = 5

1702. x = 7   1703. x = 8   1704. x = 8   1705. x = 2   1706. x = 8

Page 123:   Pre-Algebra Equations (Two Sides)

1707. x = 5   1708. x = 5   1709. x = 2   1710. x = 7   1711. x = 3

1712. x = 8   1713. x = 7   1714. x = 3   1715. x = 9   1716. x = 9

Page 124:   Pre-Algebra Equations (Two Sides)

1717. x = 4   1718. x = 6   1719. x = 7   1720. x = 9   1721. x = 8

1722. x = 9   1723. x = 5   1724. x = 8   1725. x = 8   1726. x = 2

Page 125: Pre-Algebra Equations (Two Sides)

1727. $x = 6$ 1728. $x = 5$ 1729. $x = 7$ 1730. $x = 9$ 1731. $x = 7$

1732. $x = 2$ 1733. $x = 5$ 1734. $x = 9$ 1735. $x = 3$ 1736. $x = 7$

Page 126: Simplifying Expressions

1737. $15x - 7$ 1738. $4x$ 1739. $2x$ 1740. $4x + 1$

1741. $7x + 3$ 1742. $8x$ 1743. $7x + 4$

Page 127: Simplifying Expressions

1744. $2x$ 1745. $7x$ 1746. 2 1747. $5x$

1748. $-5x + 13$ 1749. $-12x + 55$ 1750. $-49x + 44$

Page 128: Simplifying Expressions

1751. $20x + 14$ 1752. 1 1753. $-3x$ 1754. $11x$

1755. $-9x$ 1756. $72x + 57$ 1757. $-12x - 7$

Page 129: Simplifying Expressions

1758. $5x + 13$ 1759. $16x + 14$ 1760. $-2x$ 1761. $-5x$

1762. $-x + 1$ 1763. $6x$ 1764. $14x + 6$

Page 130: Simplifying Expressions

1765. $2x$ 1766. $2x + 1$ 1767. $-8x + 1$ 1768. $8x + 2$

1769. 14 1770. $-5x$ 1771. $3x - 11$

Page 131: Simplifying Expressions

1772. $-6x - 7$ 1773. $6x - 11$ 1774. $-x + 9$ 1775. $16x + 16$

1776. $x - 10$ 1777. $16x + 12$ 1778. $8x + 1$

Page 132: Simplifying Expressions

1779. $-14x$ 1780. $4x - 1$ 1781. $x - 16$ 1782. $8x + 2$

1783. $6x + 1$ 1784. $6x$ 1785. $-11x + 8$

Page 133: Simplifying Expressions

1786. $9x - 1$ 1787. $9x$ 1788. $-10x + 5$ 1789. $5x$

1790. $6x + 6$ 1791. $6x + 4$ 1792. $-9x + 8$

Page 134:   Simplifying Expressions

1793. $-5x - 13$ 1794. $3x$ 1795. $3x + 7$ 1796. $30x - 41$

1797. $-12x - 2$ 1798. $4x + 3$ 1799. $12x + 15$

Page 135:   Simplifying Expressions

1800. $10x + 3$ 1801. $5x + 7$ 1802. $4x + 7$ 1803. $-7x$

1804. $-3x + 16$ 1805. $-11x + 10$ 1806. $6x$

Page 136:   Simplifying Expressions

1807. $-3x + 14$ 1808. $2x + 1$ 1809. $-6x + 20$ 1810. $-5x + 14$

1811. $-2x$ 1812. $2x$ 1813. $9x - 8$

Page 137:   Simplifying Expressions

1814. $8x + 1$ 1815. $-5x - 15$ 1816. $-8x - 3$ 1817. $-8x + 7$

1818. $7x + 5$ 1819. 0 1820. $8x$

Page 138:   Inequalities - Addition and Subtraction

1821. $x \leq 5$ 1822. $x \leq 15$ 1823. $x \geq 8$ 1824. $x < -4$ 1825. $x < -2$

1826. $x \geq 8$

Page 139:   Inequalities - Addition and Subtraction

1827. $x > -6$ 1828. $x < 16$ 1829. $x \leq -2$ 1830. $x \leq -7$ 1831. $x < 6$

1832. $x \geq 6$

Page 140:   Inequalities - Addition and Subtraction

1833. $x \geq 0$ 1834. $x \geq 13$ 1835. $x < 4$ 1836. $x > 16$ 1837. $x \geq -7$

1838. $x \leq -1$

Page 141:   Inequalities - Multiplication and Division

1839. $x > 30$ 1840. $x \leq 4/3$ 1841. $x \geq 5/6$ 1842. $x \geq 6$ 1843. $x \leq 8$

1844. $x \leq 6/5$

Page 142:   Inequalities - Multiplication and Division

1845. $x \geq 1/3$ 1846. $x \geq 30$ 1847. $x < 1$ 1848. $x \leq 1/3$ 1849. $x > 6$

1850. $x \leq 5/4$

Page 143: Inequalities - Multiplication and Division

1851. $x > 35$ 1852. $x < 3/2$ 1853. $x > 2/3$ 1854. $x > 7$ 1855. $x < 12$

1856. $x > 5/4$

Page 144: Find the Area and Perimeter

1857. P=58 A=132   1858. P=32 A=49   1859. P=46 A=91

1860. P=36 A=72   1861. P=52 A=129   1862. P=34 A=45

Page 145: Find the Area and Perimeter

1863. P=30 A=31   1864. P=53 A=119   1865. P=60 A=165

1866. P=55 A=126   1867. P=48 A=100   1868. P=48 A=108.43

Page 146: Find the Area and Perimeter

1869. P=62 A=141   1870. P=32 A=42   1871. P=32 A=37.68

1872. P=54 A=91   1873. P=49 A=104   1874. P=40 A=63

Page 147: Find the Volume and Surface Area

1875. V=14 cm³ cm³ SA=28 cm² cm²

1876. V=254.47 cm³ cm³ SA=226 cm² cm²

1877. V=33 cm³ cm³ SA=64 cm² cm²

1878. V=192.42 cm³ cm³ SA=187 cm² cm²

1879. V=210 cm³ cm³ SA=214 cm² cm²

1880. V=34 cm³ cm³ SA=50 cm² cm²

Page 148: Find the Volume and Surface Area

1881. V=180 cm³ cm³ SA=154 cm² cm²

1882. V=157 cm³ cm³ SA=201 cm² cm²

1883. V=34 cm³ cm³ SA=50 cm² cm²

1884. V=222 cm³ cm³ SA=249 cm² cm²

1885. V=5 cm³ cm³ SA=19 cm² cm²

1886. V=75 cm³ cm³ SA=110 cm² cm²

Page 149:   Find the Volume and Surface Area

1887. V=268 cm³ cm³   SA=201 cm² cm²

1888. V=192 cm³ cm³   SA=230.4 cm² cm²

1889. V=524 cm³ cm³   SA=314 cm² cm²

1890. V=9.42 cm³ cm³   SA=25 cm² cm²

1891. V=70 cm³ cm³   SA=124.3 cm² cm²

1892. V=191 cm³ cm³   SA=206 cm² cm²

Page 150:   Calculate the area of each circle.

1893. A=803.84 cm²   1894. A=907.46 cm²   1895. A=113.04 cm²

1896. A=200.96 cm²   1897. A=1,133.54 cm²   1898. A=530.66 cm²

Page 151:   Calculate the area of each circle.

1899. A=153.86 cm²   1900. A=12.56 cm²   1901. A=530.66 cm²

1902. A=706.50 cm²   1903. A=907.46 cm²   1904. A=314.00 cm²

Page 152:   Calculate the area of each circle.

1905. A=803.84 cm²   1906. A=3.14 cm²   1907. A=113.04 cm²

1908. A=379.94 cm²   1909. A=1,256.00 cm²   1910. A=1,017.36 cm²

Page 153:   Calculate the circumference of each circle.

1911. C=125.60 cm   1912. C=113.04 cm   1913. C=18.84 cm

1914. C=31.40 cm   1915. C=75.36 cm   1916. C=69.08 cm

Page 154:   Calculate the circumference of each circle.

1917. C=94.20 cm   1918. C=31.40 cm   1919. C=56.52 cm

1920. C=12.56 cm   1921. C=50.24 cm   1922. C=43.96 cm

Page 155:   Calculate the circumference of each circle.

1923. C=113.04 cm   1924. C=25.12 cm   1925. C=119.32 cm

1926. C=62.80 cm   1927. C=69.08 cm   1928. C=50.24 cm

Page 156:   Measure of Center - Mean

1929. Mean = 54   1930. Mean = 49.5   1931. Mean = 67

1932. Mean = 47.333   1933. Mean = 46.833   1934. Mean = 60.833

Page 157:   Measure of Center - Mean

1935. Mean = 46.556   1936. Mean = 42.714   1937. Mean = 46.125

1938. Mean = 49.778   1939. Mean = 53.25   1940. Mean = 34

Page 158:   Measure of Center - Median

1941. Median = 53.5   1942. Median = 52.5   1943. Median = 56

1944. Median = 83   1945. Median = 35   1946. Median = 52

Page 159:   Measure of Center - Median

1947. Median = 59   1948. Median = 45   1949. Median = 65.5

1950. Median = 44   1951. Median = 34   1952. Median = 88

Page 160:   Measure of Center - Mode

1953. Mode = none   1954. Mode = none   1955. Mode = none

1956. Mode = none   1957. Mode = 29, 64   1958. Mode = 80

Page 161:   Measure of Center - Mode

1959. Mode = none   1960. Mode = none   1961. Mode = none

1962. Mode = 81   1963. Mode = none   1964. Mode = none

Page 162:   Measure of Variability - Range

1965. Range = 81   1966. Range = 86   1967. Range = 92   1968. Range = 75

1969. Range = 57   1970. Range = 76

Page 163:   Measure of Variability - Range

1971. Range = 63   1972. Range = 66   1973. Range = 94   1974. Range = 84

1975. Range = 80   1976. Range = 82

Made in the USA
Las Vegas, NV
12 April 2024

88544213R00111